10 Texas STAAR Grade 7 Math Practice Tests

The Ultimate Test Prep Collection with Answer Explanations

Dr. A. Nazari

10 Practice Tests

Grade 7 Math — Engineered for Mastery

Welcome to **The Architect's Workshop.**

Ten practice tests. Ten floors of a tower you're about to build. Each test lays another level of understanding — from foundation to capstone.

- **Foundation (Tests 1–3):** Learn the blueprints
- **Framework (Tests 4–6):** Build structural strength
- **Finishing (Tests 7–9):** Refine under pressure
- **Capstone (Test 10):** The final inspection

Precision. Practice. Perfection.

> **❝** Every great structure
> is built one floor at a time.
> Ten levels of practice makes
> your math rock-solid. **❞**

✚ The Architectural Plan ✚

A 4-phase construction plan for complete mastery

Phase I: Foundation (Tests 1–3)

Untimed. Explore the test format and question types. Read every answer explanation after each test. Goal: understand the blueprint before you build.

Phase II: Framework (Tests 4–6)

Add a timer (70 minutes). Focus on the topics that gave you trouble in Phase I. Practice showing complete work and labeling units. Goal: build structural strength.

Phase III: Finishing (Tests 7–9)

Full timed conditions (60 minutes). Simulate the real exam. Review only the questions you missed — don't re-study what you already know. Goal: refine under real pressure.

Phase IV: Capstone (Test 10)

Your final inspection. Full exam conditions. This is the capstone — compare with Test 1 and measure your entire growth arc. Goal: prove your mastery.

📋 Blueprint Specifications

✚ **Multiple Choice** — select the single best answer

✚ **Multi-Select** — choose ALL that apply

✚ **Short Answer** — show your process

✚ **Open Response** — explain and justify

Engineering Principles

Precision techniques for building every solution

The D.R.A.F.T. Method

D — **Define** — What exactly is the question asking? Write it in your own words.

R — **Retrieve** — Pull out given information. List numbers, units, and relationships.

A — **Assemble** — Choose the right formula or strategy. Set up the equation or proportion.

F — **Figure** — Compute step by step. Show every operation. Label units.

T — **Test** — Does the answer make sense? Re-read the question. Verify the units.

Seven Precision Rules

1. **Read twice, solve once.** The first read gives context; the second reveals what to compute.

3. **Estimate before calculating.** A quick mental approximation catches major errors.

5. **Track your signs.** Rational number operations are the #1 error source in Grade 7.

7. **Never submit blanks.** On open response, even a partial setup can earn credit.

Timing Blueprint

Tests 1–3: **Untimed** *(learn the blueprints)* > *Tests 4–6:* **70 min** *(build speed)* > *Tests 7–10:* **60 min** *(exam conditions)*

Every great structure starts with a solid plan. Study the blueprints, learn from each test, and build something extraordinary.

Prepare your station before every construction session

Required Equipment

✏️	**Precision Pencils**	Two or more #2 pencils, kept sharp. A dull tool produces dull work.
🧽	**Quality Eraser**	Clean, smudge-free corrections. Architects revise — so will you.
📝	**Scratch Paper**	Your drafting workspace. All calculations happen here first.
📐	**Ruler / Protractor**	Precision tools for scale drawings, angle measures, and geometry.
⏱️	**Timer**	Begin using from Phase II (Test 4) onward.
🔇	**Quiet Workspace**	A clean, well-lit station free from distractions.

Approved Materials

- ✔ Pencil and eraser
- ✔ Scratch paper (provided on test day)
- ✔ Ruler or protractor (if specified)
- ✔ Reference sheet (if provided)

Restricted Materials

- ✖ Calculator (unless specified)
- ✖ Electronic devices
- ✖ Notes or reference materials
- ✖ Communication with others

Ten practice tests provide the most comprehensive preparation available. The 4-phase structure (Foundation → Framework → Finishing → Capstone) ensures skills build progressively and durably.

Keys to a successful build:

- *Space tests 2–3 days apart — never more than one per day*
- *Phases I–II: review answers together and discuss strategies*
- *Phases III–IV: let them work independently, then debrief results*
- *Compare Test 1 with Test 10 to celebrate the full construction arc*

Find more at
ViewMath.com/TX-Grade7

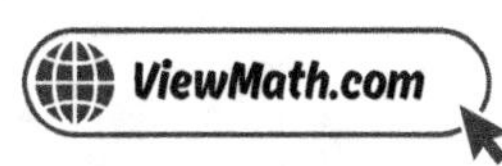

Math Reference Sheet

Symbol	Name	What It Means	
$(\ )$	Parentheses	Do this part first.	$(3+4) \times 2 = 14$
10^3	Exponent	Multiply the base by itself that many times. $10^3 = 1{,}000$	
$\frac{a}{b}$	Fraction	a parts out of b equal parts; also means $a \div b$.	
$\frac{7}{3}$	Improper Fraction	Numerator $\geq$ denominator.	$\frac{7}{3} = 2\frac{1}{3}$
0.45	Decimal	A number with a decimal point.	$0.45 = \frac{45}{100}$
$> < =$	Comparison	Greater than, less than, equal to.	$0.5 > 0.35$
$(3,5)$	Ordered Pair	A point on the coordinate plane: (x, y).	

Key Formulas

- **Volume of a rectangular prism:**

  $V = l \times w \times h$

- **Order of operations:**

  Parentheses $\rightarrow$ Exponents $\rightarrow$ Multiply/Divide $\rightarrow$ Add/Subtract

- **Powers of 10:**

  $10^1 = 10 \quad 10^2 = 100$

  $10^3 = 1{,}000 \quad 10^4 = 10{,}000$

- **Fraction as division:**

  $\frac{a}{b} = a \div b$

⊞ Place Value Chart

Millions	1,000,000	**Decimals**	
Hundred-Thousands	100,000	Tenths	0.1
Ten-Thousands	10,000	Hundredths	0.01
Thousands	1,000	Thousandths	0.001
Hundreds	100		
Tens	10		
Ones	1		

Each place is **10**× the place to its right, and $\frac{1}{10}$ of the place to its left.

📖 Key Math Vocabulary

- **Sum** — the result of addition
- **Difference** — the result of subtraction
- **Product** — the result of multiplication
- **Quotient** — the result of division
- **Remainder** — what's left over after dividing
- **Factor** — a number you multiply
- **Expression** — numbers and operations without =
- **Equation** — a math sentence with =

- **Numerator** — the top number of a fraction
- **Denominator** — the bottom number of a fraction
- **Mixed number** — a whole number + a fraction
- **Equivalent fractions** — fractions with equal value
- **Decimal** — a number written with a decimal point
- **Volume** — the space inside a 3D shape
- **Coordinate plane** — a grid with x and y axes
- **Ordered pair** — (x, y) location on the plane

🔍 Word Problem Clue Words

- **Add** (+): total, altogether, combined, sum, increase, more than
- **Subtract** (−): difference, how many more, fewer, remain, decrease, left

- **Multiply** (×): each, every, groups of, times, product, per, of (with fractions)
- **Divide** (÷): share equally, split, each group, how many groups, quotient, per

Find more at
ViewMath.com/TX-Grade7

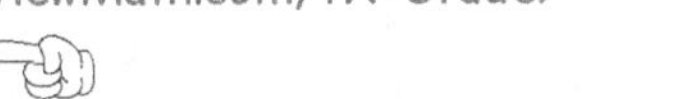

▦ Multiplication Table ▦

You may use this table during your practice tests!

×	1	2	3	4	5	6	7	8	9	10	11
1	1	2	3	4	5	6	7	8	9	10	11
2	2	4	6	8	10	12	14	16	18	20	22
3	3	6	9	12	15	18	21	24	27	30	33
4	4	8	12	16	20	24	28	32	36	40	44
5	5	10	15	20	25	30	35	40	45	50	55
6	6	12	18	24	30	36	42	48	54	60	66
7	7	14	21	28	35	42	49	56	63	70	77
8	8	16	24	32	40	48	56	64	72	80	88
9	9	18	27	36	45	54	63	72	81	90	99
10	10	20	30	40	50	60	70	80	90	100	110
11	11	22	33	44	55	66	77	88	99	110	121

💡 How to Use This Table

To find **4 × 7**:

1. Find **4** in the left column (blue).
2. Find **7** in the top row (blue).
3. Follow the row and column until they meet: the answer is **28**!

Tip: You can also use this table for division! If you know 28 ÷ 4 = ?, find 28 in the 4's row. The column header gives you the answer: **7**!

🏢 Construction Log 🏢

Architect: _______________________________   **Project Start:** _______________

Floor 1   Date: __________   Score: _____ / _____   %: _____   ★★★★★
Floor 2   Date: __________   Score: _____ / _____   %: _____   ★★★★★
Floor 3   Date: __________   Score: _____ / _____   %: _____   ★★★★★

Floor 4   Date: __________   Score: _____ / _____   %: _____   ★★★★★
Floor 5   Date: __________   Score: _____ / _____   %: _____   ★★★★★
Floor 6   Date: __________   Score: _____ / _____   %: _____   ★★★★★

Floor 7   Date: __________   Score: _____ / _____   %: _____   ★★★★★
Floor 8   Date: __________   Score: _____ / _____   %: _____   ★★★★★
Floor 9   Date: __________   Score: _____ / _____   %: _____   ★★★★★

Floor 10   Date: __________   Score: _____ / _____   %: _____   ★★★★★

Certification Level: Apprentice (0–39%)   Technician (40–59%)   Engineer (60–79%)   **Master Architect (80–100%)**

Record your certification level after each test!

Under Construction

____

____

____

____

Construction Complete

★ ____

★ ____

★ ____

★ ____

Find more at
ViewMath.com/TX-Grade7

★ Table of Contents ★

Here's what we'll explore together!

⭐ Practice Test 1 3

⭐ Practice Test 2 13

⭐ Practice Test 3 23

⭐ Practice Test 4 33

⭐ Practice Test 5 43

⭐ Practice Test 6 51

⭐ Practice Test 7 59

⭐ Practice Test 8 69

⭐ Practice Test 9 79

⭐ Practice Test 10 89

⭐ Answer Key & Explanations 100

 Let's learn and have fun!

Practice Test 1

 30 Questions

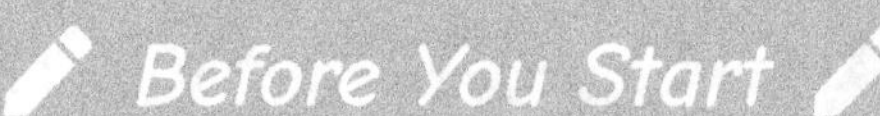

✏️ Before You Start ✏️

- ✓ **Read each question carefully** before choosing your answer.
- ✓ **Show your work** on scratch paper when you need to.
- ✓ **Skip hard questions** and come back to them later.
- ✓ **Check your answers** when you're done.
- ✓ **Take your time** — there's no rush!

⭐ You've Got This! ⭐

Do your best and show what you know!

1. The table shows a proportional relationship. Find k and the missing value.

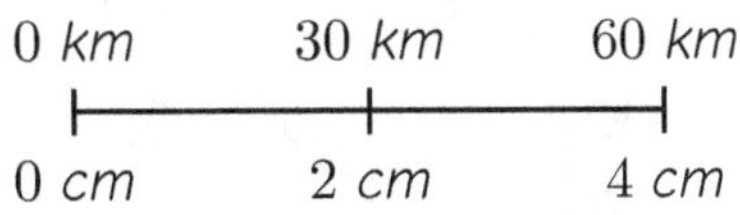

x	y
3	7.5
?	15
10	25

Your Answer:

2. A map uses the scale shown below. If two cities are 6.5 cm apart on the map, what is the actual distance?

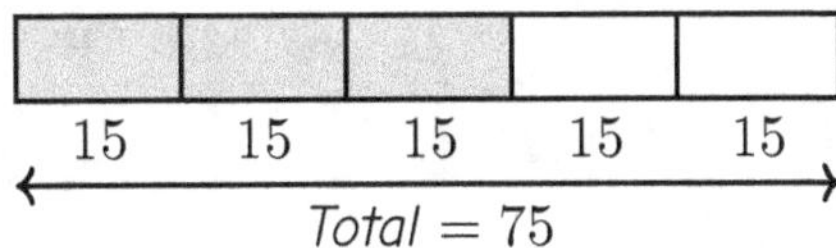

(A) 90 km

(B) 95 km

(C) 97.5 km

(D) 100 km

3. The tape diagram below represents a whole quantity. What percent of the total is the shaded portion?

(A) 40%

(B) 45%

(C) 50%

(D) 60%

4. *A plant was 8 inches tall. It grew to 14 inches. What is the percent increase?*

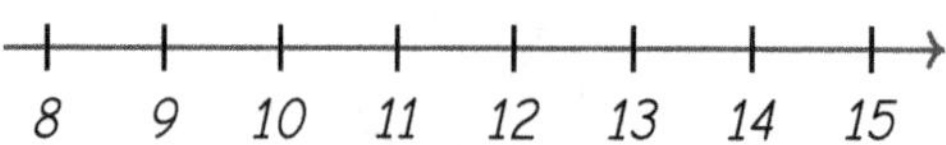

(A) 43%

(B) 57%

(C) 60%

(D) 75%

5. *The bar graph shows the interest earned on three different deposits over 2 years. Which deposit had the highest interest rate?*

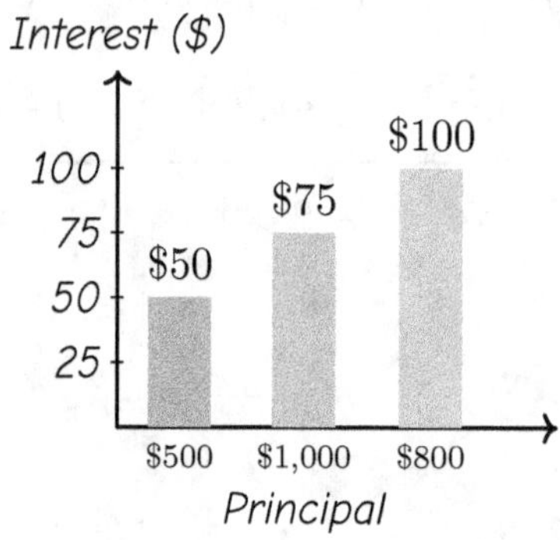

(A) Deposit of $500 (earned $50)

(B) Deposit of $1,000 (earned $75)

(C) Deposit of $800 (earned $100)

(D) All have the same rate

6. *A credit card charges 24% annual interest. You owe $500 for 6 months. How much interest do you owe?*

Your Answer:

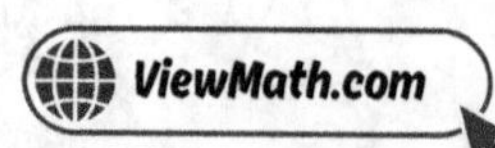

7. *The bar graph shows a family's monthly budget. Which category is a variable expense?*

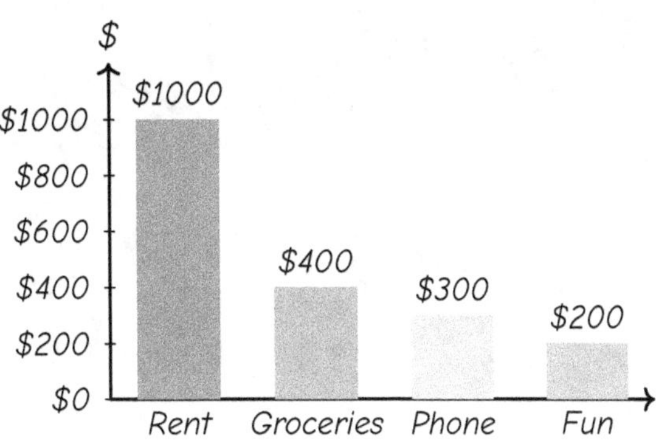

(A) Rent

(B) Phone

(C) Groceries

(D) All of them

8. *You invest $1,000 at 6% compounded yearly. After 2 years, the total is:*

(A) $1,120

(B) $1,123.60

(C) $1,060

(D) $1,126

9. *Evaluate $\frac{x^2-1}{4}$ when $x = 5$.*

Your Answer:

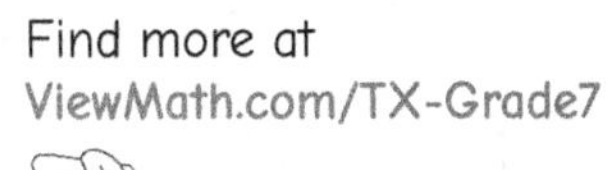
Find more at
ViewMath.com/TX-Grade7

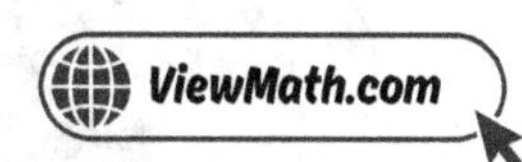

10. Simplify $\frac{1}{2}x + \frac{3}{4}x - 2$.

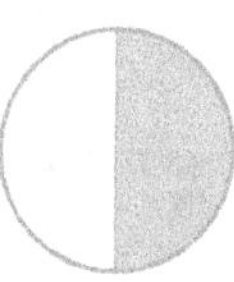

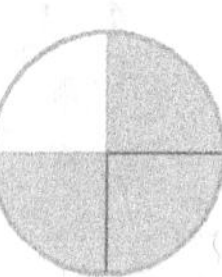

Your Answer:

11. Look at the number line. The expression $3(x+2)$ equals the distance from A to B. What is x?

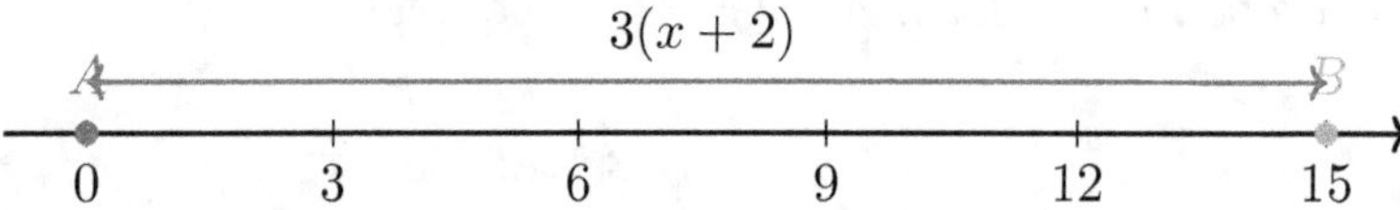

(A) $x = 5$ (B) $x = 7$

(C) $x = 3$ (D) $x = 13$

12. A number line shows a point at x. The equation $\dfrac{x}{4} + 1.5 = 5$ tells you how to find x. Solve and identify where x is on the number line.

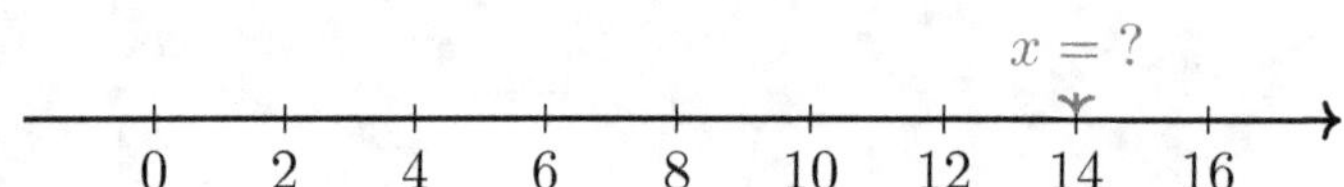

Your Answer:

13. You earn $9 per hour and need more than $100 for a purchase. You already have $28. Write and solve an inequality for the number of hours h you need to work.

Your Answer:

14. What type of circle is used to graph $x > 3$ on a number line?

(A) Closed circle at 3

(B) Open circle at 3

(C) Closed circle at 0

(D) Open circle at 0

15. A drawing uses $1\ cm = 7\ m$. You redraw at $1\ cm = 7\ m$ (the same scale). A line is 5 cm long. How long is it on the new drawing?

Your Answer:

16. You know two sides of a triangle are 7 cm and 10 cm, and the angle between them is 45°. How many different triangles can you draw?

(A) None

(B) Exactly one

(C) Exactly two

(D) Infinitely many

17. A triangle has sides 11, 13, and 20. Does it satisfy the triangle inequality?

Your Answer:

18. What shape is the cross-section when a cone is sliced vertically through its apex?

Your Answer:

Find more at
ViewMath.com/TX-Grade7

19. *Are all circles similar to each other?*

(A) No, they have different radii

(B) No, only congruent circles are similar

(C) Yes, all circles are similar

(D) Only if they have the same center

20. *A pentagon can be divided into a rectangle 6 cm by 4 cm and a triangle with base 6 cm and height 2 cm. What is the total area?*

(A) $24\ cm^2$

(B) $30\ cm^2$

(C) $36\ cm^2$

(D) $48\ cm^2$

21. *A triangular prism has a triangular base with base 6 cm and height 4 cm, and the prism is 10 cm long. The three rectangular faces have widths 6 cm, 5 cm, and 5 cm. What is the total surface area?*

(A) $184\ cm^2$

(B) $160\ cm^2$

(C) $124\ cm^2$

(D) $120\ cm^2$

22. *A cube has a volume of $64\ cm^3$. What is the side length?*

(A) 2 cm

(B) 4 cm

(C) 8 cm

(D) 16 cm

23. *A company wants to know if customers like a new product. They survey only customers who bought the product last week. What is the problem with this sample?*

(A) The sample is too large

(B) The sample is biased toward recent buyers

(C) The sample is random

(D) There is no problem

Find more at
ViewMath.com/TX-Grade7

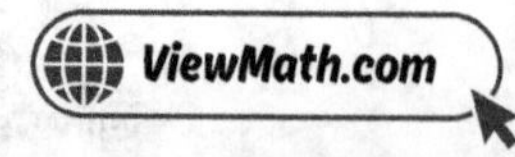

24. A sample of 40 oranges from a farm has a mean weight of 200 grams. If the farm ships 10,000 oranges, predict the total weight.

(A) 200,000 grams

(B) 2,000,000 grams

(C) 8,000 grams

(D) 400,000 grams

25. The MAD for data set $\{4, 6, 8, 10, 12\}$ with mean 8 is:

(A) 2

(B) 2.4

(C) 3

(D) 8

26. Use the stem-and-leaf plot to answer the question.

Stem	Leaf
1	5 8
2	0 3 3 7
3	1 5 9
4	2

Key: 1 | 5 means 15

How many data values are greater than 25?

(A) 3

(B) 4

(C) 5

(D) 6

27. A class of 40 students: 16 prefer dogs, 12 prefer cats, 8 prefer fish, 4 prefer birds. What percent prefer dogs?

(A) 16%

(B) 25%

(C) 40%

(D) 80%

Find more at
ViewMath.com/TX-Grade7

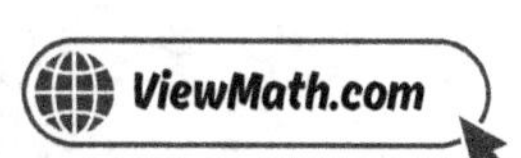

28. *A probability model has outcomes X, Y, and Z. $P(X) = 0.15$ and $P(Y) = 0.45$. What is $P(Z)$?*

Your Answer:

29. *A spinner has 4 equal sections $(1, 2, 3, 4)$ and a die is rolled. What is the probability that both show a 3?*

(A) $\frac{1}{10}$

(B) $\frac{1}{24}$

(C) $\frac{2}{10}$

(D) $\frac{1}{4}$

30. *A spinner has a 60% chance of landing on blue. In a simulation of 80 spins, how many times would you expect blue to appear?*

Your Answer:

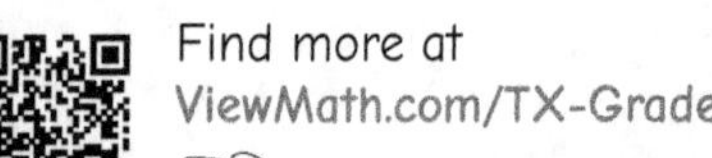

Find more at
ViewMath.com/TX-Grade7

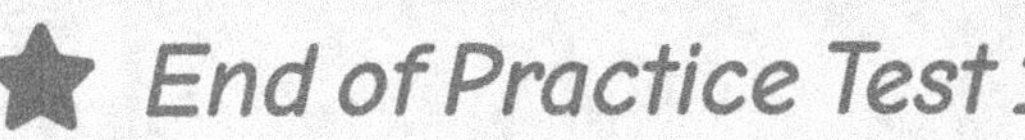

⭐ End of Practice Test 1 ⭐

Great job finishing the test!

📋 My Score

I got _____________ out of 30 questions right.

*Check your answers in the **Answer Key** at the back of the book.*

💡 *Review any questions you missed. That's how we learn!*

📊 Check Your Score Online!

Visit **ViewMath Academy** to enter your answers and see which topics you need to review. You can also explore lessons, take quizzes, track your scores, and save your progress!

viewmath.com/score/7.1.TX.16

Or go to viewmath.com/score and enter code: 7.1.TX.16

Practice Test 2

 30 Questions

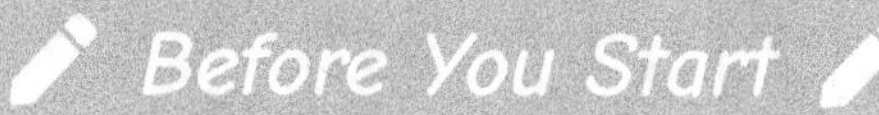

✏️ Before You Start ✏️

- ✔ **Read each question carefully** before choosing your answer.
- ✔ **Show your work** on scratch paper when you need to.
- ✔ **Skip hard questions** and come back to them later.
- ✔ **Check your answers** when you're done.
- ✔ **Take your time** — there's no rush!

⭐ You've Got This! ⭐

Do your best and show what you know!

1. *The table below shows a proportional relationship between gallons of gas and miles driven. What is k and what does it mean?*

Gallons (x)	Miles (y)
2	54
5	135
8	216

(A) $k = 2$; uses 2 gallons per trip

(B) $k = 27$; drives 27 miles per gallon

(C) $k = 54$; drives 54 miles on 2 gallons

(D) $k = 135$; drives 135 miles on 5 gallons

2. *The double number line below relates inches on a blueprint to actual feet. A hallway measures 3.5 inches on the blueprint. What is the actual length?*

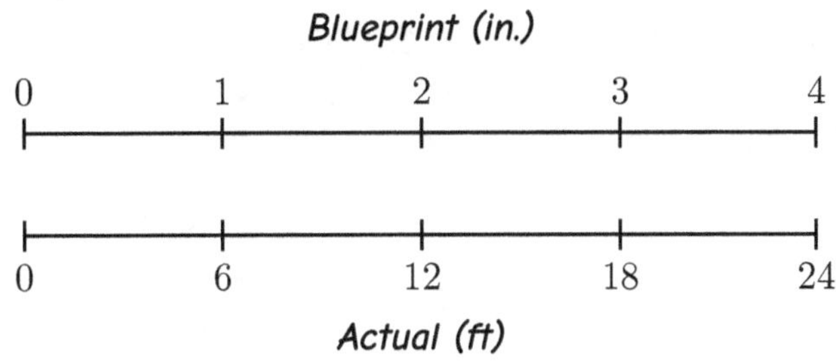

Your Answer:

3. *The circle graph shows how 200 students get to school. How many students take the bus?*

Your Answer:

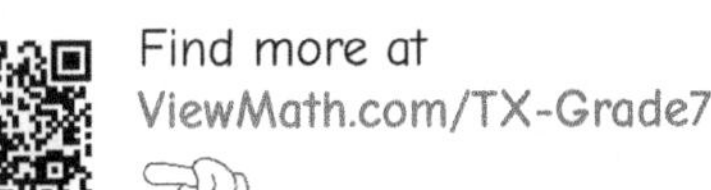

Find more at
ViewMath.com/TX-Grade7

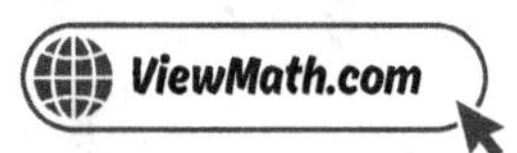

4. What multiplier represents a 25% increase?

(A) 0.25   (B) 0.75

(C) 1.25   (D) 25

5. What is the simple interest on $900 at 8% for 6 months?

(A) $36   (B) $72

(C) $108   (D) $432

6. You owe $250 on a credit card with 20% annual interest. If you pay nothing for 1 year, how much will you owe?

(A) $270   (B) $300

(C) $275   (D) $350

7. Which of the following is a fixed expense?

(A) Groceries   (B) Movie tickets

(C) Monthly rent   (D) Clothing

8. Use $A = P(1 + r)^t$ to find the value of $1,500 at 6% compounded yearly for 2 years.

Your Answer

9. What is the value of $\frac{x}{2} - 4$ when $x = 10$?

(A) 3   (B) -1

(C) 1   (D) 9

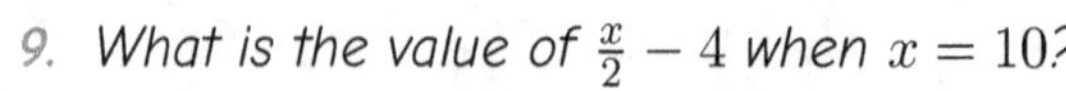

Find more at
ViewMath.com/TX-Grade7

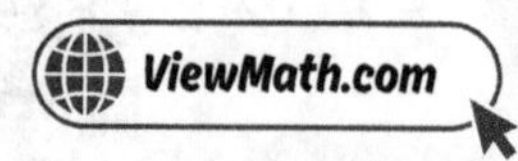

10. Simplify $-2a + 5b + 7a - 3b$.

Your Answer:

11. Three friends split the cost of a gift equally. Each person also pays $4 for wrapping. Each person pays $16 total. What is the cost c of the gift?

(A) $c = 48$

(B) $c = 36$

(C) $c = 12$

(D) $c = 60$

12. Solve $-0.3x + 4.2 = 1.8$.

Your Answer:

13. The table shows ticket prices for groups. A school has $200 to spend and must also pay a $35 bus fee. Write and solve an inequality to find the maximum number of students s who can go.

Group Size	Price per Student
1–10	$12
11–25	$9
26+	$7

Assume the group qualifies for the $9 rate.

Your Answer:

14. What inequality is shown by a closed circle at -6 with shading to the right?

Your Answer:

Find more at
ViewMath.com/TX-Grade7

15. When you redraw a scale drawing at a smaller scale (more real units per cm), the new drawing is:

(A) Larger than the original drawing

(B) Smaller than the original drawing

(C) The same size as the original drawing

(D) A different shape than the original drawing

16. You know all three angles of a triangle: 50°, 60°, and 70°. How many triangles can be drawn?

(A) None

(B) Exactly one

(C) Exactly three

(D) More than one (infinitely many)

17. Two angles of a triangle are 35° and 75°. What is the third angle?

(A) 60°

(B) 70°

(C) 80°

(D) 110°

18. A cube can be sliced to produce a hexagonal cross-section. How?

(A) By cutting parallel to the base

(B) By cutting diagonally through all six faces

(C) By cutting vertically through the center

(D) It is impossible to get a hexagon from a cube

19. A tree casts a 16-ft shadow. A 4-ft fence post casts a 5-ft shadow at the same time. How tall is the tree? Round to the nearest tenth.

Your Answer:

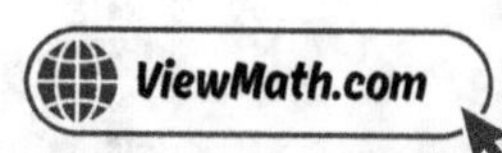

20. A shape is made of a square with side 8 cm and a semicircle with diameter 8 cm attached to one side. What is the area? Use $\pi \approx 3.14$.

(A) 64 cm^2

(B) 89.12 cm^2

(C) 114.24 cm^2

(D) 139.12 cm^2

21. A paint company needs to paint the outside of a box that is 3 m by 2 m by 1 m. Each can of paint covers 5 m^2. How many cans are needed?

(A) 3

(B) 4

(C) 5

(D) 6

22. A planter box is 3 ft long, 1.5 ft wide, and 2 ft deep. How many cubic feet of soil are needed to fill it?

Your Answer:

23. A random sample means that:

(A) Only the best members are chosen

(B) The largest possible group is chosen

(C) Every member of the population has an equal chance of being selected

(D) Only volunteers are selected

24. A sample of 80 customers found that $\frac{3}{8}$ prefer online shopping. If the store has 2,400 customers, predict how many prefer online shopping.

Your Answer:

25. The box plots show quiz scores (out of 20) for two classes.

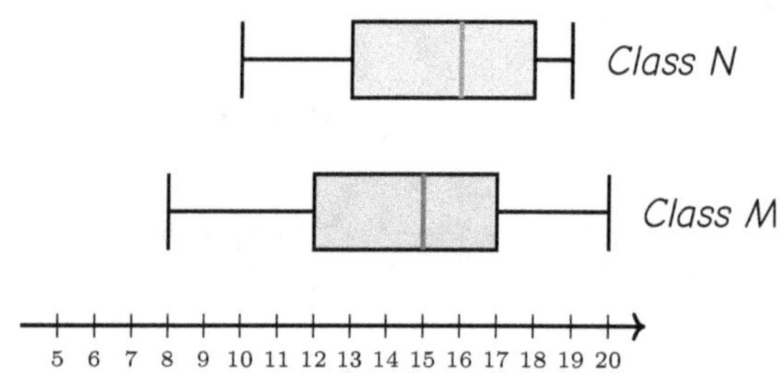

What is the IQR for each class?

(A) Class M: 5, Class N: 5

(B) Class M: 12, Class N: 9

(C) Class M: 5, Class N: 9

(D) Class M: 12, Class N: 5

26. The stem-and-leaf plot shows the ages of people at a family reunion.

Stem	Leaf
0	5 8 9
1	2 4 6
2	0 5 8
3	2 5 7 9
4	1 5
5	0 8
6	3 7
7	2

Key: 3 | 2 means 32 years old

Find the range and median of the ages.

Your Answer:

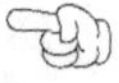
Find more at
ViewMath.com/TX-Grade7

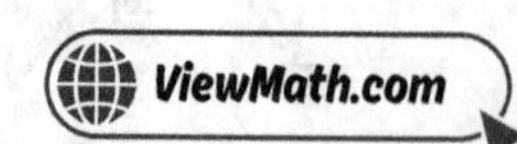

27. *A circle graph represents:*

(A) *How data changes over time*   (B) *How a whole is divided into parts*

(C) *The range of a data set*   (D) *Individual data points*

28. *A student creates a probability model for a spinner.* $P(red) = 0.25$, $P(blue) = 0.25$, $P(green) = 0.25$, $P(yellow) = 0.25$. *After* 200 *spins, the results are:* red $= 55$, blue $= 48$, green $= 52$, yellow $= 45$. *Is the model a good fit?*

(A) *No, the model is completely wrong*   (B) *Yes, the observed frequencies are reasonably close to* 50 *each*

(C) *No, because the results are not exactly* 50 *each*   (D) *Yes, because* 200 *is divisible by* 4

29. *The tree diagram shows the outcomes for flipping two coins. What is the probability of getting at least one tail?*

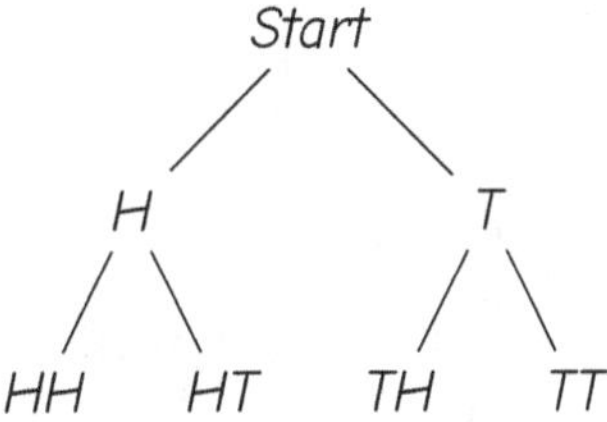

(A) $\frac{1}{4}$ (B) $\frac{1}{2}$

(C) $\frac{3}{4}$ (D) $\frac{2}{4}$

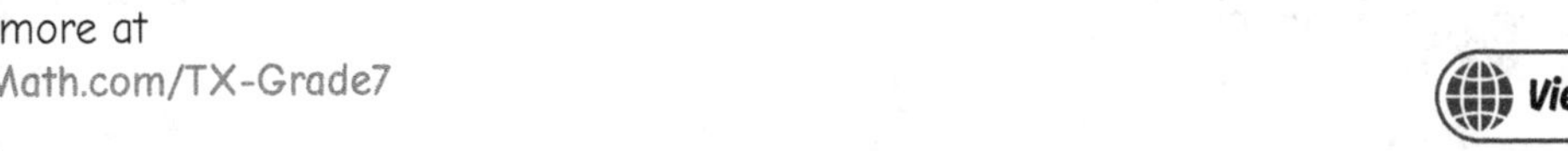

30. The bar graph shows the results of simulating a spinner 100 times. The spinner is supposed to land on each color with equal probability. Based on the simulation, does the spinner appear fair?

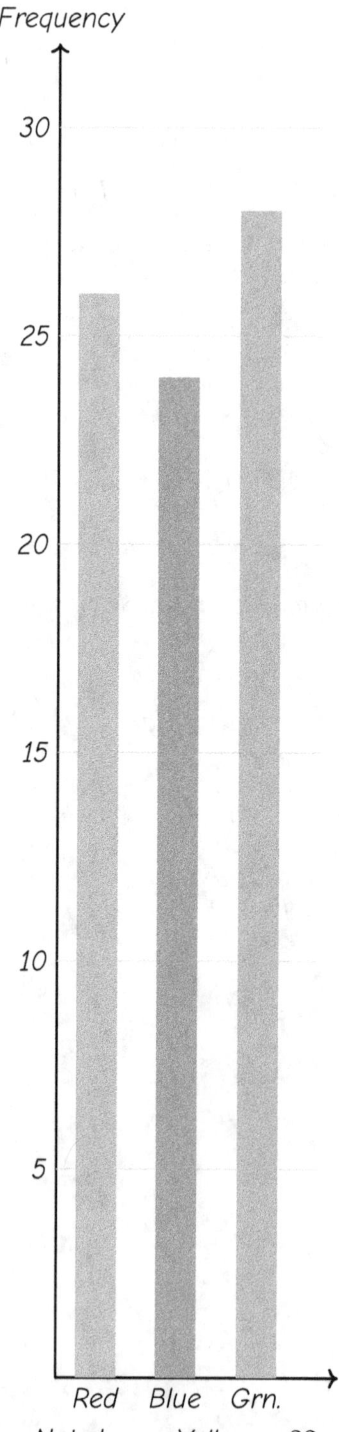

Not shown: Yellow = 22

(A) No, because the bars are different heights

(B) Yes, the frequencies $(26, 24, 28, 22)$ are all roughly close to 25

(C) Green is clearly the most likely color

(D) Cannot be determined from the graph

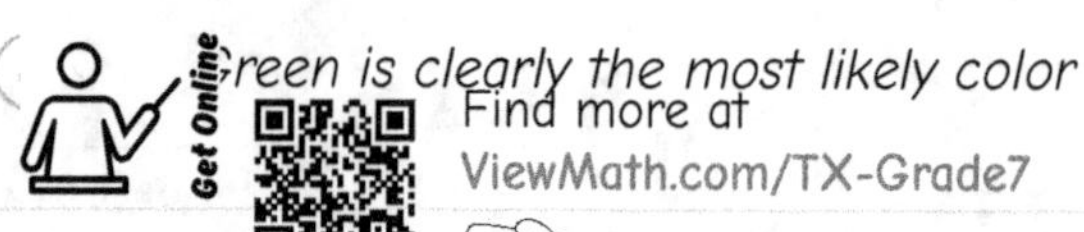

 # End of Practice Test 2 

Great job finishing the test!

My Score

I got _____________ out of 30 questions right.

*Check your answers in the **Answer Key** at the back of the book.*

Review any questions you missed. That's how we learn!

📊 Check Your Score Online!

Visit **ViewMath Academy** to enter your answers and see which topics you need to review. You can also explore lessons, take quizzes, track your scores, and save your progress!

viewmath.com/score/7.1.TX.17

Or go to viewmath.com/score and enter code: 7.1.TX.17

Practice Test 3

 30 Questions

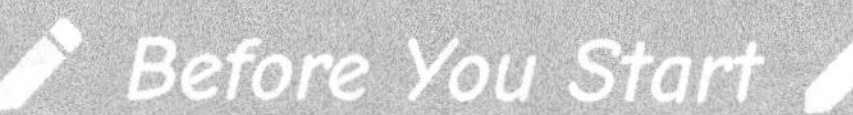

✏️ Before You Start ✏️

- ✓ **Read each question carefully** before choosing your answer.
- ✓ **Show your work** on scratch paper when you need to.
- ✓ **Skip hard questions** and come back to them later.
- ✓ **Check your answers** when you're done.
- ✓ **Take your time** — there's no rush!

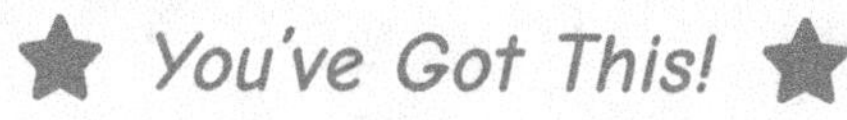

⭐ You've Got This! ⭐

Do your best and show what you know!

1. A proportional table has $k = \frac{3}{4}$. Which pair could be in the table?

   (A) $(8, 10)$ (B) $(8, 6)$

   (C) $(12, 8)$ (D) $(4, 2)$

2. A recipe for 4 people uses $\frac{3}{2}$ cups of rice. How much rice is needed for 10 people?

   (A) $2\frac{1}{2}$ cups (B) 3 cups

   (C) $3\frac{3}{4}$ cups (D) 5 cups

3. 63 is 90% of what number?

   (A) 56.7 (B) 67

   (C) 70 (D) 72

4. A town's population was 12,000 last year and is 13,200 this year. What is the percent increase?

   (A) 8% (B) 10%

   (C) 12% (D) 15%

5. What is the simple interest on $1,200 at 3.5% for 2 years?

   (A) $42 (B) $72

   (C) $84 (D) $210

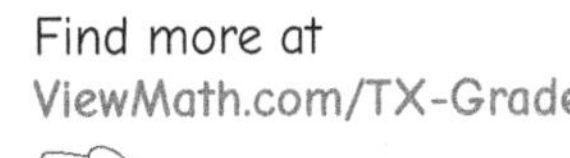

Find more at
ViewMath.com/TX-Grade7

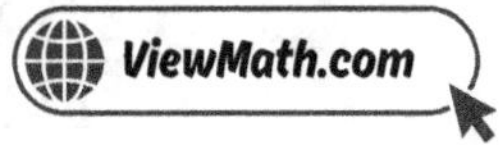

6. You charge $300 on a credit card with an 18% annual interest rate. If you pay it off after 1 year, how much interest do you owe?

(A) $18

(B) $36

(C) $54

(D) $318

7. Your entertainment budget is $80/month. After 5 months, you've spent $350 total on entertainment. Are you under or over budget?

(A) Under budget by $50

(B) Over budget by $50

(C) Under budget by $30

(D) Over budget by $30

8. You deposit $500 at 10% compounded yearly. What is the total after 2 years?

(A) $600

(B) $605

(C) $550

(D) $610

9. What is the value of $x^2 + 3x$ when $x = -2$?

(A) 10

(B) -2

(C) -10

(D) 2

10. Simplify $-6y + 4y - y$.

(A) $-3y$

(B) $-y$

(C) $3y$

(D) $-11y$

Find more at
ViewMath.com/TX-Grade7

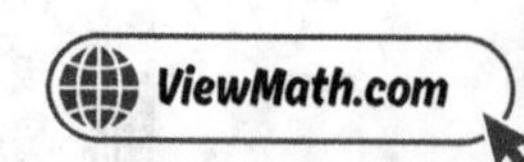

11. A student solved $3(x + 4) = 30$ and wrote: $3x + 4 = 30$, $3x = 26$, $x = \frac{26}{3}$. What was the student's error?

(A) Divided by 3 instead of subtracting

(B) Did not distribute 3 to the 4

(C) Subtracted 4 from the wrong side

(D) Forgot to check the answer

12. Solve $2.5x - 7.5 = 10$.

Your Answer:

13. Solve $-4x > 20$.

(A) $x > -5$

(B) $x < -5$

(C) $x > 5$

(D) $x < 5$

14. Solve $-5x + 15 \geq 30$. Describe the graph.

Your Answer:

15. Original scale: $1\ cm = 10\ m$. New scale: $1\ cm = 40\ m$. What is the scale ratio?

(A) $\frac{1}{4}$

(B) 4

(C) 30

(D) $\frac{1}{30}$

16. A triangle has side lengths 9, 12, and 15. What type of triangle is it?

(A) Equilateral

(B) Isosceles

(C) Right

(D) It cannot be formed

Find more at
ViewMath.com/TX-Grade7

ViewMath.com

17. *The table below shows four sets of triangle conditions. Which one produces a unique triangle?*

Set	Given Information	Condition
A	Angles: 50°, 60°, 70°	AAA
B	Sides: 3, 4, 5	SSS
C	Angles: 45°, 90°, 45°	AAA
D	One side: 8 cm	—

(A) Set A

(B) Set B

(C) Set C

(D) Set D

18. *You cut a cone with a vertical slice through its apex (tip). What shape is the cross-section?*

(A) Circle

(B) Triangle

(C) Rectangle

(D) Semicircle

19. *A 5-ft child stands next to a wall that casts a 15-ft shadow. The child casts a 3-ft shadow. How tall is the wall?*

(A) 9 ft

(B) 25 ft

(C) 45 ft

(D) 10 ft

20. *A T-shaped figure can be broken into a 12 cm by 2 cm horizontal rectangle and a 2 cm by 8 cm vertical rectangle. What is the total area?*

(A) 24 cm^2

(B) 32 cm^2

(C) 40 cm^2

(D) 48 cm^2

Find more at
ViewMath.com/TX-Grade7

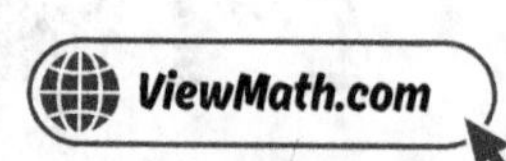

21. If all dimensions of a rectangular prism are doubled, what happens to its surface area?

(A) It doubles

(B) It triples

(C) It quadruples

(D) It increases by a factor of 8

22. A composite solid is made of two rectangular prisms joined together as shown. What is the total volume?

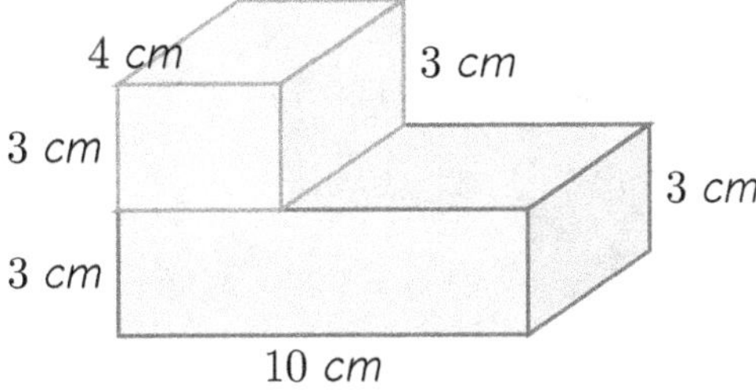

Your Answer:

23. A poll found that 35% of a random sample of 400 students prefer pizza for lunch. Which statement is the best inference?

(A) Exactly 35% of all students prefer pizza

(B) About 35% of the entire student population likely prefers pizza

(C) No students prefer anything other than pizza

(D) The poll is invalid because not all students were surveyed

24. A sample of 20 plants in a greenhouse has a mean height of 18 cm. A sample of 80 plants has a mean height of 16 cm. Which estimate is probably more reliable?

(A) The sample of 20 because it has a higher mean

(B) The sample of 80 because it is larger

(C) Both are equally reliable

(D) Neither is reliable

25. *Class A: median 85, IQR 8, range 20. Class B: median 80, IQR 15, range 40. A student says "Class B is better because it has a wider range." Is the student correct?*

(A) *Yes — wider range means more talent*

(B) *No — wider range means less consistency, and Class A has a higher median*

(C) *Yes — Class B's scores spread more, which is always better*

(D) *No — range does not measure anything useful*

26. *How many data values are shown in this stem-and-leaf plot?*

Stem 5: 1 3 5   *Stem 6:* 0 2 4 7 8   *Stem 7:* 1 6   *Stem 8:* 0

Your Answer

27. *Two circle graphs compare how Class A and Class B spend their free time. Class A shows 45% on sports; Class B shows 30% on sports. If Class A has 20 students and Class B has 30 students, how many more students in Class B play sports than in Class A?*

(A) *0 — the same number*

(B) *9 is more in Class B*

(C) *3 is more in Class A*

(D) *1 more in Class A*

28. *A bag contains 4 red marbles and 1 blue marble. Which type of probability model does this represent?*

(A) *Uniform*

(B) *Non-uniform*

(C) *Both uniform and non-uniform*

(D) *Neither*

29. *Two dice are rolled. What is the probability of getting a sum of 2?*

(A) $\frac{1}{6}$

(B) $\frac{1}{12}$

(C) $\frac{1}{36}$

(D) $\frac{2}{36}$

Find more at
ViewMath.com/TX-Grade7

30. You simulate flipping a coin 3 times and counting how many times all 3 are heads. In 40 simulations, all 3 heads occurs 6 times. What is the experimental probability?

(A) $\frac{6}{40} = 0.15$

(B) $\frac{1}{8} = 0.125$

(C) $\frac{3}{40} = 0.075$

(D) $\frac{6}{120} = 0.05$

Find more at
ViewMath.com/TX-Grade7

 # End of Practice Test 3 

Great job finishing the test!

 My Score

I got ___________ out of 30 questions right.

*Check your answers in the **Answer Key** at the back of the book.*

Review any questions you missed. That's how we learn!

Check Your Score Online!

Visit **ViewMath Academy** to enter your answers and see which topics you need to review. You can also explore lessons, take quizzes, track your scores, and save your progress!

viewmath.com/score/7.1.TX.18

Or go to viewmath.com/score and enter code: 7.1.TX.18

Practice Test 4

📋 30 Questions

✏️ Before You Start ✏️

- ✓ **Read each question carefully** before choosing your answer.
- ✓ **Show your work** on scratch paper when you need to.
- ✓ **Skip hard questions** and come back to them later.
- ✓ **Check your answers** when you're done.
- ✓ **Take your time** — there's no rush!

⭐ You've Got This! ⭐

Do your best and show what you know!

1. What is the constant of proportionality (k) for the table below?

x	y
3	12
5	20
8	32

(A) 3

(B) 4

(C) 8

(D) 12

2. A truck delivers 540 packages in 3 trips. How many trips are needed to deliver 1,620 packages?

Your Answer:

3. 9 is 12% of what number?

(A) 1.08

(B) 36

(C) 72

(D) 75

4. A pond had 400 fish last year and 340 this year. What is the percent decrease?

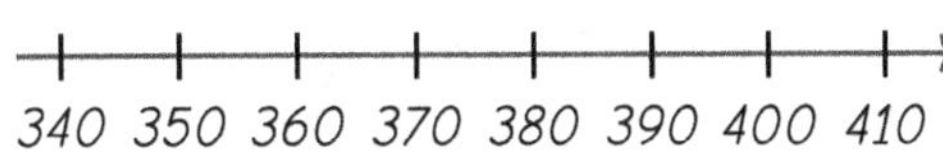

Your Answer:

5. Find the simple interest: $P = \$500$, $r = 6\%$, $t = 3$ years.

(A) \$30

(B) \$60

(C) \$90

(D) \$150

Find more at
ViewMath.com/TX-Grade7

6. A credit union is most similar to which of the following?

(A) A department store

(B) A bank

(C) A stock market

(D) A government office

7. Which formula correctly represents the relationship between income, expenses, and savings?

(A) Income = Savings + Expenses

(B) Savings = Expenses − Income

(C) Income = Expenses − Savings

(D) Expenses = Income + Savings

8. \$900 at 10% compounded yearly for 3 years. What is the total amount?

Your Answer:

9. Which expression represents "the product of 7 and a number p, decreased by 12"?

(A) $7(p - 12)$

(B) $12 - 7p$

(C) $7p - 12$

(D) $7 + p - 12$

10. Simplify $7x + 3x$.

(A) $10x$

(B) $21x$

(C) $10x^2$

(D) $73x$

11. Solve $7(x - 2) = 0$.

(A) $x = 0$

(B) $x = -2$

(C) $x = 2$

(D) $x = 7$

12. Solve $\dfrac{x}{2} + \dfrac{x}{4} = 9$.

Your Answer:

13. Look at the number line. Which inequality has this solution?

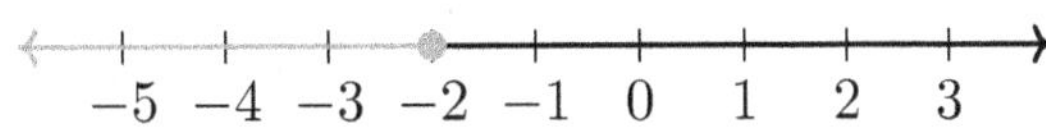

$$-5 \quad -4 \quad -3 \quad -2 \quad -1 \quad 0 \quad 1 \quad 2 \quad 3$$

(A) $3x + 1 > -5$

(B) $-4x \geq 8$

(C) $2x - 3 < -7$

(D) $x + 5 > 3$

14. Solve $10 - 4x < 2$. Describe the graph.

Your Answer:

15. A map uses $1\ cm = 20\ km$. You redraw it at $1\ cm = 10\ km$. What happens to the drawing?

(A) Every length is halved

(B) Every length stays the same

(C) Every length is doubled

(D) Every length is multiplied by 10

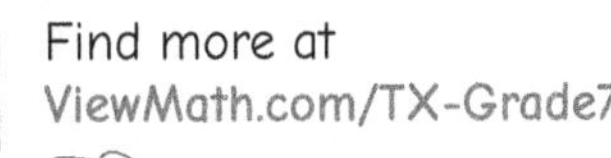

Find more at
ViewMath.com/TX-Grade7

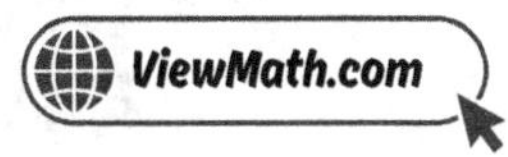

16. The four rectangles below all have the same perimeter of 16 cm but different dimensions. What does this tell us?

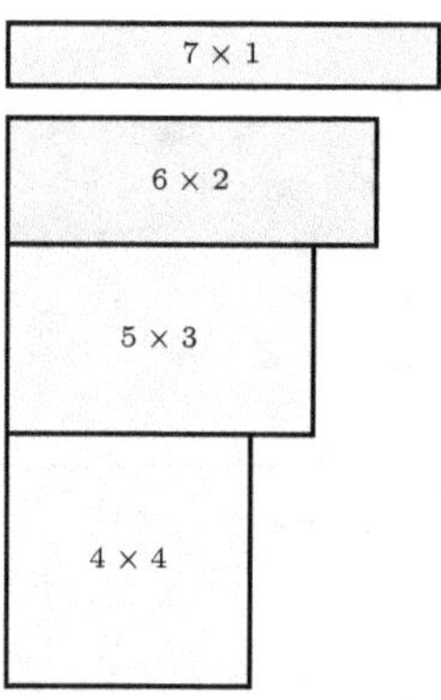

(A) A given perimeter determines exactly one rectangle

(B) A given perimeter allows more than one rectangle

(C) All rectangles with the same perimeter have the same area

(D) It is impossible for multiple rectangles to share a perimeter

17. Sides 5, 12, 13. Is this a right triangle?

(A) No, because $5 + 12 \neq 13$

(B) No, because the sides are not equal

(C) Yes, because $5^2 + 12^2 = 13^2$

(D) Yes, because $5 + 12 = 17 > 13$

18. A hexagonal prism is sliced parallel to its base. What is the cross-section?

(A) Rectangle

(B) Triangle

(C) Hexagon

(D) Circle

19. Triangle $LMN \sim$ Triangle XYZ. $LM = 10$, $MN = 14$, $LN = 16$, $XY = 5$. Find YZ.

(A) 7

(B) 8

(C) 9

(D) 28

Find more at
ViewMath.com/TX-Grade7

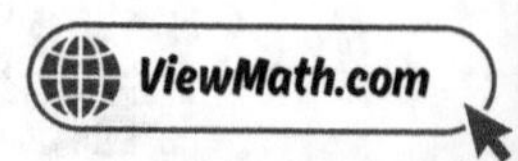

20. A banner is shaped like a rectangle 18 in by 6 in with a triangle cut from one end. The triangle has a base of 6 in and a height of 4 in. What is the area of the banner?

(A) 84 in^2

(B) 96 in^2

(C) 108 in^2

(D) 120 in^2

21. How many faces does a rectangular prism have?

(A) 4

(B) 5

(C) 6

(D) 8

22. What unit is used to measure volume?

(A) Square units (cm^2)

(B) Cubic units (cm^3)

(C) Linear units (cm)

(D) Degrees

23. The diagram below shows four different sampling methods used to survey students at a school. Which method is most likely to produce an unbiased sample?

Method A
Survey the football team

Method B
Survey every 10th student on the school roster

Method C
Post survey online and wait for responses

Method D
Survey students in one math class

(A) Method A

(B) Method B

(C) Method C

(D) Method D

Find more at
ViewMath.com/TX-Grade7

24. A sample of 40 students found that 10 prefer vanilla ice cream. If the school has 800 students, predict how many prefer vanilla.

Your Answer:

25. Group A has a mean of 50 and a MAD of 3. Group B has a mean of 60 and a MAD of 3. The difference in means expressed in MADs is:

(A) $\frac{10}{3} \approx 3.3$ MADs

(B) 10 MADs

(C) 3 MADs

(D) 30 MADs

26. Data: $45, 48, 52, 55, 55, 58, 61, 63$. Which is the correct stem-and-leaf plot?

(A) Stems: $4, 5, 6$; Row 4: 5 8; Row 5: 2 5 5 8; Row 6: 1 3

(B) Stems: $4, 5, 6$; Row 4: 5 8; Row 5: 5 5 2 8; Row 6: 3 1

(C) Stems: $45, 48, 52$; no leaves

(D) Stems: $4, 5$; Row 4: 5 8; Row 5: 2 5 5 8 1 3

27. A student creates a circle graph with sectors of 100°, 80°, 60°, and 100°. Is this circle graph correct?

(A) Yes — the sectors look reasonable

(B) No — the angles add to 340°, not 360°

(C) No — there are too many sectors

(D) Yes — all angles are less than 180°

28. A coin is weighted so that heads comes up 60% of the time. What is the probability of tails?

(A) 0.60

(B) 0.50

(C) 0.40

(D) 0.30

Find more at
ViewMath.com/TX-Grade7

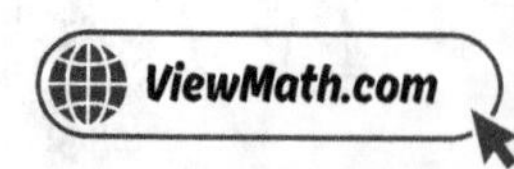

29. Look at the tree diagram for flipping three coins. What is the probability of getting exactly 2 tails? Write your answer as a fraction in simplest form.

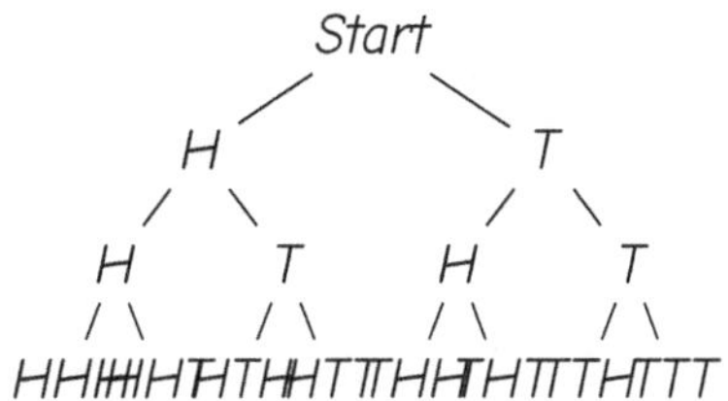

Your Answer:

30. A student wants to simulate the probability that exactly 2 out of 3 students pass a test, where each student has a 50% chance. Which model would work?

(A) Roll one die: odd = pass, even = fail

(B) Flip 3 coins: heads = pass, tails = fail; count exactly 2 heads

(C) Draw 3 marbles from a bag of 2 red and 1 blue

(D) Flip 1 coin 3 times and count all tails

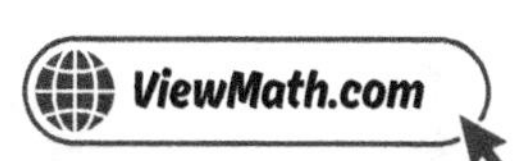

 # End of Practice Test 4 

Great job finishing the test!

 My Score

I got _____________ out of 30 questions right.

*Check your answers in the **Answer Key** at the back of the book.*

💡 *Review any questions you missed. That's how we learn!*

📊 Check Your Score Online!

Visit **ViewMath Academy** to enter your answers and see which topics you need to review. You can also explore lessons, take quizzes, track your scores, and save your progress!

viewmath.com/score/7.1.TX.19

Or go to viewmath.com/score and enter code: 7.1.TX.19

5

Practice Test 5

 *30 Questions*

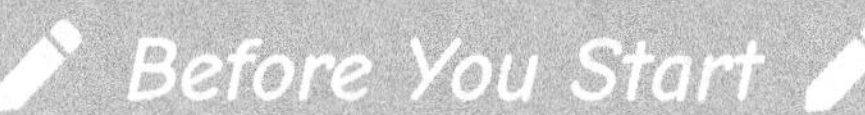

✏️ Before You Start ✏️

- ✔ **Read each question carefully** before choosing your answer.
- ✔ **Show your work** on scratch paper when you need to.
- ✔ **Skip hard questions** and come back to them later.
- ✔ **Check your answers** when you're done.
- ✔ **Take your time** — there's no rush!

⭐ You've Got This! ⭐

Do your best and show what you know!

1. A factory produces widgets at a constant rate. In $2\frac{1}{2}$ hours it makes 175 widgets. What is k in widgets per hour?

Your Answer

2. A map scale says $2\ cm = 15$ miles. Two cities are 7 cm apart on the map. What is the actual distance?

(A) 42.5 miles

(B) 45 miles

(C) 52.5 miles

(D) 105 miles

3. A school survey showed that 54 out of 200 students walk to school. What percent of students walk to school?

(A) 25%

(B) 27%

(C) 37%

(D) 54%

4. The table shows the number of visitors to a park over two months. Which month had the greatest percent change from the previous month?

Month	April	May	June
Visitors	400	500	550

(A) April to May (25% increase)

(B) April to May (20% increase)

(C) May to June (10% increase)

(D) May to June (25% increase)

5. A savings account pays 3% simple interest. You deposit $2,000. After how many years will you earn $300 in interest?

(A) 3 years

(B) 4 years

(C) 5 years

(D) 10 years

6. Name one advantage and one disadvantage of using a credit card instead of a debit card.

Your Answer:

7. You earn \$2,400 per month and save \$240. What percent of your income do you save?

(A) 24%

(B) 12%

(C) 10%

(D) 20%

8. A \$800 deposit earns \$48 in simple interest over 2 years. What is the annual interest rate?

(A) 2%

(B) 3%

(C) 4%

(D) 6%

9. Which expression means "half the sum of a number k and 10"?

(A) $\frac{k}{2} + 10$

(B) $\frac{k+10}{2}$

(C) $\frac{10k}{2}$

(D) $\frac{k}{2+10}$

10. The perimeter of a triangle is $3x + 2 + 5x - 1 + 2x + 4$. Write the simplified expression for the perimeter.

Your Answer:

11. Solve $4(x + 9) = 52$.

Your Answer:

Get Online

Find more at
ViewMath.com/TX-Grade7

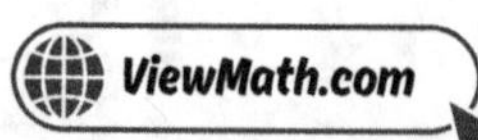

12. Which is the best first step to solve $\dfrac{x}{4} + \dfrac{1}{2} = 3$?

(A) Subtract $\frac{1}{2}$ from both sides

(B) Multiply every term by 4

(C) Divide both sides by 4

(D) Add $\frac{1}{2}$ to both sides

13. Solve $8 - 3x \geq 23$.

Your Answer:

14. Which value is NOT a solution to $x \geq -4$?

(A) $x = -4$

(B) $x = 0$

(C) $x = -5$

(D) $x = 100$

15. A floor plan uses 1 cm = 5 ft. You redraw at 1 cm = 2 ft. A room that is 3 cm by 4 cm on the original drawing. What is the area on the new drawing?

(A) 12 cm^2

(B) 30 cm^2

(C) 75 cm^2

(D) 150 cm^2

16. How many different rectangles have an area of 24 cm^2 if the sides must be whole numbers?

Your Answer:

17. Which set of measurements does NOT determine a unique triangle?

(A) Sides 3, 4, 5

(B) Sides 7 and 9 with included angle 50°

(C) Angles 40°, 70°, 70°

(D) Angles 30°, 60° with included side 10 cm

Find more at
ViewMath.com/TX-Grade7

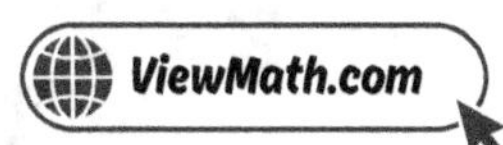
ViewMath.com

18. You slice a cone with a horizontal cut parallel to its base. What shape is the cross-section?

(A) A circle smaller than the base

(B) A circle the same size as the base

(C) A triangle

(D) A rectangle

19. Triangle ABC has sides 3, 4, 5. Triangle DEF is similar with the longest side equal to 20. What is the shortest side of triangle DEF?

(A) 8

(B) 12

(C) 15

(D) 16

20. A rectangular room is 14 ft by 10 ft. There is a built-in closet that is 3 ft by 4 ft. What is the area of the room not including the closet?

(A) 128 ft^2

(B) 140 ft^2

(C) 152 ft^2

(D) 12 ft^2

21. A cube has a surface area of 384 in^2. What is the side length?

Your Answer:

22. A rectangular fish tank is 50 cm long, 30 cm wide, and 40 cm tall. What is the volume?

(A) 120 cm^3

(B) 6,000 cm^3

(C) 60,000 cm^3

(D) 600 cm^3

Find more at
ViewMath.com/TX-Grade7

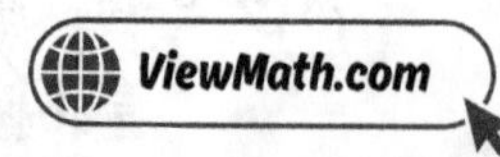

23. A doctor wants to study the effect of a new medicine on 10,000 patients. She randomly selects 500 patients for a trial. What is the sample size?

(A) 10,000

(B) 500

(C) 10,500

(D) 50

24. A random sample of 25 fish from a lake has a mean length of 14 inches. The best prediction for the mean length of all fish in the lake is:

(A) Exactly 14 inches

(B) About 14 inches

(C) More than 14 inches

(D) Less than 14 inches

25. Team X: mean 82, MAD 5. Team Y: mean 76, MAD 5. Which team performed better on average, and is the difference meaningful?

(A) Team X; no, because the difference is less than 2 MADs

(B) Team X; yes, because $\frac{6}{5} = 1.2$ MADs, which is meaningful

(C) Team X; yes, because the means are more than 2 MADs apart

(D) Team Y; yes, because its MAD is the same

26. A stem-and-leaf plot shows the row 2 | 2 2 5 8. What is the mode for values with stem 2?

(A) 22

(B) 25

(C) 28

(D) There is no mode

27. A sector represents 15% of a data set. What is its angle in degrees?

Your Answer:

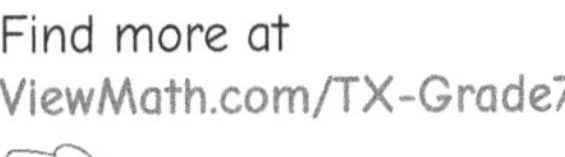
Find more at
ViewMath.com/TX-Grade7

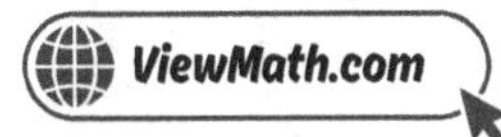

28. A bag has tiles labeled A, A, A, B, B. What is the probability of drawing a B?

(A) $\frac{1}{5}$

(B) $\frac{2}{3}$

(C) $\frac{2}{5}$

(D) $\frac{3}{5}$

29. The table shows all possible products when two dice are rolled. How many outcomes give a product of 6? Write the probability as a fraction in simplest form.

×	1	2	3	4	5	6
1	1	2	3	4	5	6
2	2	4	6	8	10	12
3	3	6	9	12	15	18
4	4	8	12	16	20	24
5	5	10	15	20	25	30
6	6	12	18	24	30	36

Your Answer

30. Which of the following is the FIRST step in designing a simulation?

(A) Record the results

(B) Run many trials

(C) Identify the event and its possible outcomes

(D) Calculate the experimental probability

Find more at
ViewMath.com/TX-Grade7

 *End of Practice Test 5* 

Great job finishing the test!

 My Score

I got _____________ out of 30 questions right.

*Check your answers in the **Answer Key** at the back of the book.*

💡 Review any questions you missed. That's how we learn!

📊 Check Your Score Online!

Visit **ViewMath Academy** to enter your answers and see which topics you need to review. You can also explore lessons, take quizzes, track your scores, and save your progress!

viewmath.com/score/7.1.TX.20

Or go to viewmath.com/score and enter code: 7.1.TX.20

6

Practice Test 6

 30 Questions

✏️ Before You Start ✏️

✔ **Read each question carefully** before choosing your answer.

✔ **Show your work** on scratch paper when you need to.

✔ **Skip hard questions** and come back to them later.

✔ **Check your answers** when you're done.

✔ **Take your time** — there's no rush!

⭐ You've Got This! ⭐

Do your best and show what you know!

1. A proportional graph passes through $(6, 15)$. What is k?

Your Answer

2. A printer prints 120 pages in 8 minutes. How long will it take to print 450 pages?

(A) 25 min

(B) 28 min

(C) 30 min

(D) 36 min

3. A library has 480 books. If 35% are fiction, how many fiction books does the library have?

(A) 148

(B) 160

(C) 168

(D) 172

4. A bike costs $200. It is on sale for 35% off. What is the sale price?

(A) $70

(B) $130

(C) $135

(D) $165

5. Two bank accounts both start with $1,000. Account A earns 5% for 2 years. Account B earns 2% for 5 years. Which earns more interest?

(A) Account A

(B) Account B

(C) They earn the same

(D) Cannot be determined

Find more at
ViewMath.com/TX-Grade7

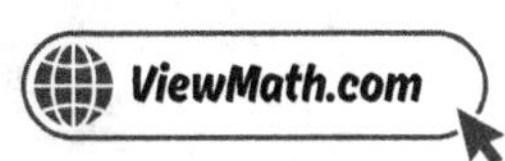

6. You buy a $500 laptop using a credit card at 12% annual interest. If you pay it all back after 1 year, what is the total cost?

A) $512

B) $560

C) $600

D) $550

7. A savings account earns 2% per year. You deposit $500 for 1 year. How much interest do you earn?

A) $2

B) $10

C) $50

D) $100

8. A $700 loan at 5% compounded yearly for 2 years. How much is owed?

A) $770

B) $771.75

C) $735

D) $772

9. Which expression means "a number n divided by 4, then increased by 3"?

A) $\frac{n+3}{4}$

B) $\frac{4}{n} + 3$

C) $\frac{n}{4} + 3$

D) $4n + 3$

10. Simplify $-4x + 9 + 7x - 3$.

A) $3x + 6$

B) $-11x + 6$

C) $3x - 6$

D) $11x + 12$

11. *Solve* $10(x - 5) = 30$.

(A) $x = 8$

(B) $x = 3$

(C) $x = -2$

(D) $x = 35$

12. *The table shows input and output values for a function. What is the rule?*

Input (x)	Output (y)
2	2.5
4	3.5
6	4.5
10	6.5

(A) $y = 0.5x + 1.5$

(B) $y = x - 0.5$

(C) $y = 0.25x + 2$

(D) $y = 2x - 1.5$

13. *Solve* $-7x - 4 < 17$.

Your Answer:

14. *A parking garage charges \$5 per hour. You have \$30. Which inequality represents the hours h you can park, and what does the graph look like?*

(A) $5h \leq 30$; *closed circle at 6, shade left from 0 to 6*

(B) $5h < 30$; *open circle at 6, shade left*

(C) $5h \geq 30$; *closed circle at 6, shade right*

(D) $5h > 30$; *open circle at 6, shade right*

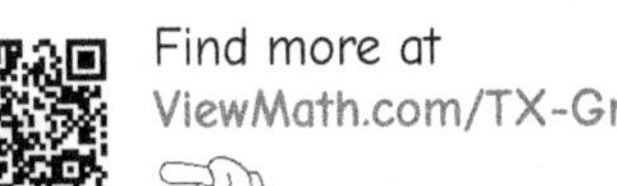

Find more at
ViewMath.com/TX-Grade7

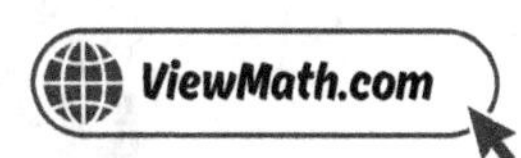

15. A blueprint uses $1\ in = 12\ ft$. You redraw at $1\ in = 4\ ft$. A door that is $0.5\ in$ wide on the original becomes how wide?

(A) 0.5 in

(B) 1 in

(C) 1.5 in

(D) 3 in

16. A circle with radius 5 cm is described as a condition. How many circles satisfy this condition?

(A) None

(B) Exactly one

(C) More than one (different positions)

(D) It depends on the center location

17. Can a triangle have sides 8, 8, and 16? Explain.

Your Answer

18. Is it possible to get a circular cross-section from a rectangular prism? Explain.

Your Answer

19. Two similar triangles have a scale factor of $\frac{2}{5}$. If a side of the larger triangle is 30 cm, what is the corresponding side of the smaller triangle?

(A) 6 cm

(B) 12 cm

(C) 15 cm

(D) 75 cm

20. A trapezoidal garden has parallel sides of 8 m and 12 m, and a height of 5 m. What is its area?

Your Answer

Find more at
ViewMath.com/TX-Grade7

21. A rectangular prism is 6 ft by 6 ft by 10 ft. What is its surface area?

(A) $72\ ft^2$

(B) $240\ ft^2$

(C) $312\ ft^2$

(D) $360\ ft^2$

22. A triangular prism has a triangular base with base 10 m and height 6 m. The prism is 8 m long. What is the volume?

Your Answer

23. A website asks visitors to rate a product. Only 5% of visitors respond. This is a:

(A) Random sample

(B) Systematic sample

(C) Voluntary response sample

(D) Stratified sample

24. Which of the following would improve the accuracy of a prediction made from a sample?

(A) Using a smaller sample size

(B) Surveying only your friends

(C) Taking multiple random samples and averaging the results

(D) Ignoring outliers in the data

25. Two groups have the same mean but different MADs. What does this tell you?

(A) One group scored higher than the other

(B) The groups have different levels of variability

(C) The groups are identical

(D) The means were calculated incorrectly

Find more at
ViewMath.com/TX-Grade7

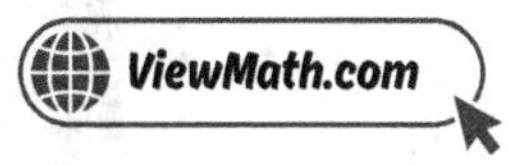

26. Which data display keeps the original data values while also showing the shape of the data?

   (A) Histogram

   (B) Box plot

   (C) Stem-and-leaf plot

   (D) Circle graph

27. A survey of 500 people found: Walking 20%, Driving 55%, Bus 15%, Other 10%. How many drive?

   (A) 55

   (B) 100

   (C) 200

   (D) 275

28. A spinner is divided into 4 sections with probabilities: red = 0.4, blue = 0.3, green = 0.2, yellow = 0.1. What type of model is this?

   (A) Uniform, because there are 4 sections

   (B) Non-uniform, because the probabilities are different

   (C) Invalid, because the probabilities don't add up to 1

   (D) Uniform, because all sections are on the same spinner

29. A coin is flipped and a die is rolled. What is the probability of getting tails and an even number?

   (A) $\frac{1}{12}$

   (B) $\frac{1}{6}$

   (C) $\frac{1}{4}$

   (D) $\frac{1}{2}$

30. What is a simulation?

   (A) A way to calculate exact probabilities

   (B) A model that uses random numbers or objects to imitate a real event

   (C) A method that always gives the same result

   (D) A graph that shows all possible outcomes

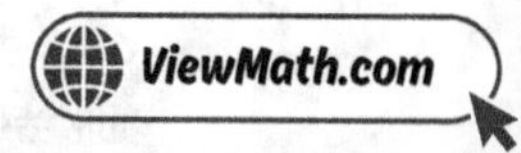

 **End of Practice Test 6** 

Great job finishing the test!

 My Score

I got _____________ out of 30 questions right.

*Check your answers in the **Answer Key** at the back of the book.*

💡 *Review any questions you missed. That's how we learn!*

📊 Check Your Score Online!

*Visit **ViewMath Academy** to enter your answers and see which topics you need to review. You can also explore lessons, take quizzes, track your scores, and save your progress!*

viewmath.com/score/7.1.TX.21

Or go to viewmath.com/score and enter code: 7.1.TX.21

Practice Test 7

 30 Questions

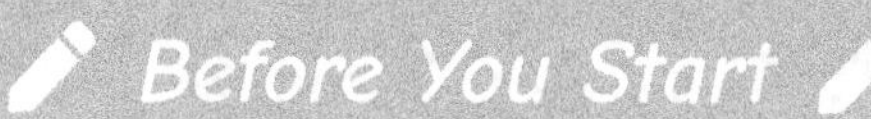

✏️ Before You Start ✏️

- ✔ **Read each question carefully** before choosing your answer.
- ✔ **Show your work** on scratch paper when you need to.
- ✔ **Skip hard questions** and come back to them later.
- ✔ **Check your answers** when you're done.
- ✔ **Take your time** — there's no rush!

⭐ You've Got This! ⭐

Do your best and show what you know!

1. Find the missing value in the proportional table.

x	y
5	20
9	?

2. A machine fills 350 bottles in 5 hours. How many bottles does it fill in 12 hours?

(A) 700

(B) 780

(C) 840

(D) 900

3. 42 is what percent of 168?

4. A jacket cost $75 last winter. This winter it costs $60. What is the percent decrease?

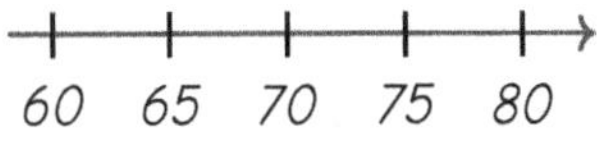

(A) 15%

(B) 20%

(C) 25%

(D) 30%

5. A loan of $5,000 at 3% per year charges $450 in interest. How many years was the loan?

Find more at
ViewMath.com/TX-Grade7

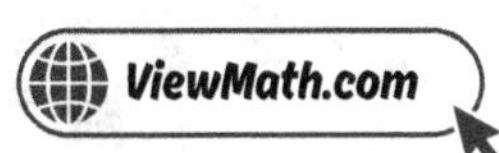

6. Which type of card lets you spend money you already have in your bank account?

(A) Credit card

(B) Gift card

(C) Debit card

(D) Rewards card

7. You earn \$3,200 per month. Your goal is to save 15% of income. How many dollars should you save each month?

Your Answer

8. Which grows faster over time?

(A) Simple interest

(B) Compound interest

(C) They grow at the same rate

(D) It depends on the principal

9. Which expression represents "6 more than twice a number n"?

(A) $6n + 2$

(B) $2n + 6$

(C) $2(n + 6)$

(D) $6 \times 2n$

10. Simplify $\frac{3}{4}n - \frac{1}{4}n + 5$.

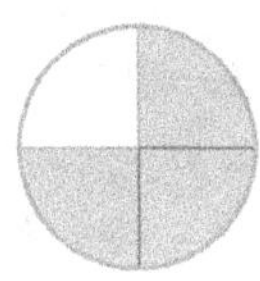

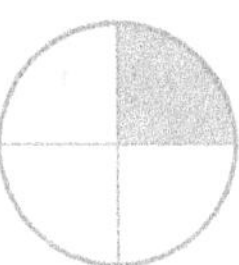

(A) $\frac{1}{2}n + 5$

(B) $n + 5$

(C) $\frac{3}{4}n + 5$

(D) $\frac{2}{4}n - 5$

11. Each side of an equilateral triangle is $(2x + 5)$ cm. The perimeter is 57 cm. Find x.

Your Answer:

12. Solve $-1.5x + 4.5 = -3$.

(A) $x = -5$

(B) $x = 1$

(C) $x = 5$

(D) $x = -1$

13. Solve $\dfrac{x}{3} + 4 < 10$.

(A) $x < 18$

(B) $x < 42$

(C) $x < 2$

(D) $x < 6$

Get Online

Find more at
ViewMath.com/TX-Grade7

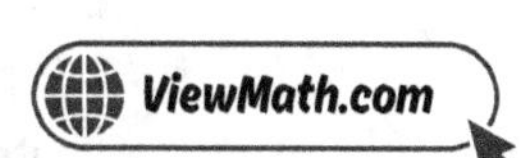

14. Which of the following is the correct way to read the graph: open circle at 0, shading to the left?

(A) All positive numbers

(B) All negative numbers

(C) All numbers less than 0

(D) All numbers less than or equal to 0

15. A drawing uses $1\ cm = 12\ m$. You redraw at $1\ cm = 4\ m$. A hallway is 5 cm on the original. How long is it on the new drawing?

Your Answer:

16. Can you draw a triangle with sides 5 cm, 5 cm, and 5 cm?

(A) No, it is impossible

(B) Yes, exactly one triangle

(C) Yes, more than one triangle

(D) Yes, but only if one angle is 90°

17. A triangle has sides 6 cm, 8 cm, and 10 cm. How many triangles can be constructed with these measurements?

(A) None

(B) Exactly one

(C) Exactly two

(D) Infinitely many

18. Which statement about cross-sections is true?

(A) The cross-section of any 3D figure is always a circle

(B) The cross-section depends on the angle and position of the cut

(C) A cross-section is always the same shape as the base

(D) Only prisms can have cross-sections

Find more at
ViewMath.com/TX-Grade7

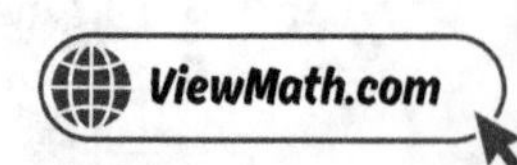

19. A photo is 4 inches by 6 inches. It is enlarged so the longer side is 15 inches. What is the shorter side of the enlarged photo?

(A) 8 in

(B) 10 in

(C) 12 in

(D) 13 in

20. An H-shaped figure is made of three rectangles. The top and bottom bars are each 10 cm by 2 cm. The middle vertical bar is 2 cm by 6 cm. What is the total area?

Your Answer:

21. A cereal box is 30 cm tall, 20 cm wide, and 6 cm deep. What is its surface area?

(A) $1{,}440\ cm^2$

(B) $1{,}560\ cm^2$

(C) $1{,}800\ cm^2$

(D) $3{,}600\ cm^2$

22. A storage container is 2 ft long, 1.5 ft wide, and 3 ft tall. How many cubic feet can it hold?

(A) $6\ ft^3$

(B) $6.5\ ft^3$

(C) $9\ ft^3$

(D) $10\ ft^3$

23. A random sample of 80 out of 2,000 students found that 20 students ride the bus. Predict how many students in the school ride the bus.

Your Answer:

Find more at
ViewMath.com/TX-Grade7

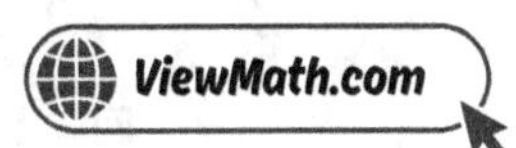

24. The bar graph shows the results of three random samples of 50 students. Each bar shows the number who said they exercise daily.

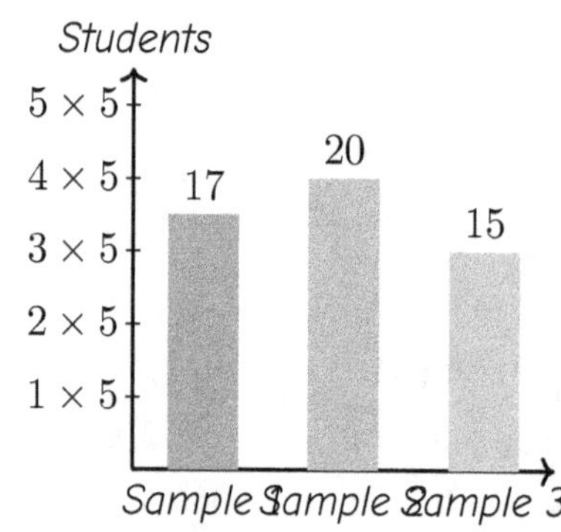

If the school has 800 students, which is the best prediction for how many exercise daily?

(A) 240

(B) 277

(C) 320

(D) 340

25. Explain one advantage of using IQR instead of range to describe the spread of a data set.

Your Answer:

26. Why is it important to include a key (e.g., "3 | 4 means 34") in a stem-and-leaf plot?

(A) It makes the plot look nicer

(B) It tells the reader how to interpret the stems and leaves

(C) It is not important

(D) It replaces the title

27. A survey of 120 students asked about their favorite after-school activity. The results are:

Activity	Students
Sports	48
Music	30
Art	18
Reading	24

Find the percent and angle (in degrees) for each category.

Your Answer:

28. Which of the following is an example of a uniform probability model?

(A) A bag with 3 red, 5 blue, and 2 green marbles   (B) A fair number cube with 6 sides

(C) A spinner with sections of different sizes   (D) A weighted coin

29. A coin is flipped and a die is rolled. What is the probability of getting heads and a number greater than 4?

(A) $\frac{1}{12}$ (B) $\frac{1}{6}$

(C) $\frac{2}{12}$ (D) $\frac{1}{3}$

30. A game has a 25% chance of winning. Which simulation model correctly represents this?

(A) Flip a coin — heads means win   (B) Roll a die — roll a 1 means win

(C) Use a spinner with 4 equal sections — one   (D) Use digits 0–9 — digits 0–4 mean win

section means win

Find more at
ViewMath.com/TX-Grade7

 # ⭐ *End of Practice Test 7* ⭐ 

Great job finishing the test!

📋 My Score

I got _____________ out of 30 questions right.

*Check your answers in the **Answer Key** at the back of the book.*

💡 *Review any questions you missed. That's how we learn!*

📊 Check Your Score Online!

*Visit **ViewMath Academy** to enter your answers and see which topics you need to review. You can also explore lessons, take quizzes, track your scores, and save your progress!*

viewmath.com/score/7.1.TX.22

Or go to viewmath.com/score and enter code: 7.1.TX.22

Practice Test 8

☑ *30 Questions*

✏ Before You Start ✏

- ✔ **Read each question carefully** before choosing your answer.
- ✔ **Show your work** on scratch paper when you need to.
- ✔ **Skip hard questions** and come back to them later.
- ✔ **Check your answers** when you're done.
- ✔ **Take your time** — there's no rush!

★ You've Got This! ★

Do your best and show what you know!

1. A car travels at a constant speed. It goes 150 miles in 2.5 hours. What is the constant of proportionality?

(A) 30 mph   (B) 60 mph

(C) 75 mph   (D) 150 mph

2. The table shows how many batches of muffins a bakery can make and the total flour used. How many cups of flour are needed for 15 batches?

Batches	Flour (cups)
3	$7\frac{1}{2}$
6	15
9	$22\frac{1}{2}$
15	?

Your Answer:

3. 56 is 80% of what number?

Your Answer:

4. A gym membership increased from \$45 per month to \$54 per month. What is the percent increase?

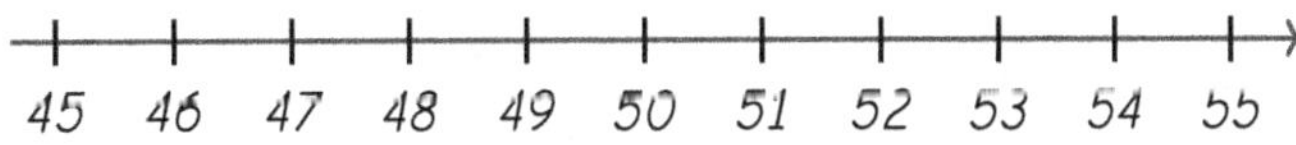

(A) 9%   (B) 15%

(C) 18%   (D) 20%

5. You earned $180 in interest on $2,000 over 3 years. What was the interest rate?

Your Answer:

6. The bar graph shows interest earned on a $1,000 deposit at three different banks after 1 year. Which bank pays the highest interest rate?

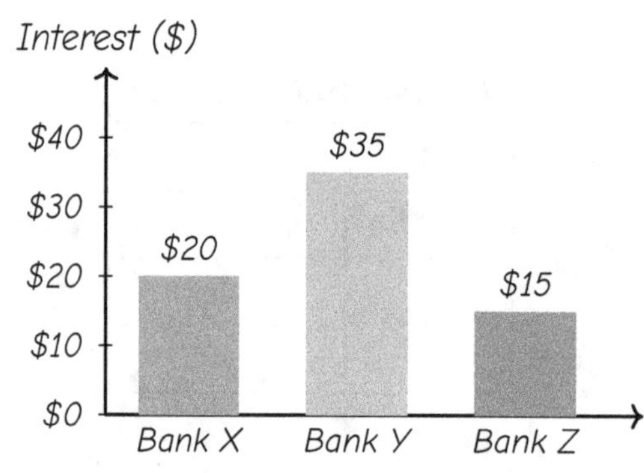

(A) Bank X

(B) Bank Y

(C) Bank Z

(D) They all pay the same

7. A student earns $200 per month from a part-time job. She spends $60 on a phone plan (fixed) and $80 on food (variable). How much can she save?

(A) $140

(B) $60

(C) $80

(D) $120

8. In the formula $A = P(1 + r)^t$, what does t represent?

(A) Tax rate

(B) Total amount

(C) Number of years

(D) Type of account

9. Evaluate $-4y + 10$ when $y = 3$.

(A) 22

(B) -2

(C) 2

(D) -22

10. Write an expression with three terms that simplifies to $4n + 7$.

Your Answer:

11. The diagram shows two equal groups. Which equation represents this, and what is x?

$$\left.\begin{array}{c} \boxed{x} + \boxed{4} \\ \text{Group 1} \end{array} \quad \begin{array}{c} \boxed{x} + \boxed{4} \\ \text{Group 2} \end{array}\right\} = 22$$

(A) $2(x + 4) = 22,\ x = 7$

(B) $2x + 4 = 22,\ x = 9$

(C) $x + 8 = 22,\ x = 14$

(D) $2(x + 4) = 22,\ x = 3$

12. Solve $\dfrac{x}{6} - \dfrac{2}{3} = \dfrac{1}{2}$.

(A) $x = 3$

(B) $x = 7$

(C) $x = -1$

(D) $x = 10$

13. Solve $\dfrac{x}{2} + 7 \leq 12$.

Your Answer:

14. A student has test scores of 78, 85, and 90. She needs an average of at least 85 on four tests to earn an A. Solve the inequality to find the minimum score s on the fourth test, and describe the graph.

$$\xrightarrow{\ \ |\ \ |\ \ |\ \ |\ \ |\ \ |\ \ |\ \ |\ \ }$$
$$65\ 70\ 75\ 80\ 85\ 90\ 95\ 100$$

Your Answer:

15. A floor plan uses 1 cm = 6 ft. On this plan, a room is 9 cm long. On a second plan, the same room is 3 cm long. What scale does the second plan use?

Your Answer:

16. The shortest two sides of a triangle are 7 and 9. What is the largest whole-number length the third side can be?

Your Answer:

17. What condition (SSS, SAS, ASA, AAS, or AAA) gives infinitely many triangles?

Your Answer:

18. Which 3D figure always has circular cross-sections when cut parallel to its base?

(A) Rectangular prism

(B) Triangular pyramid

(C) Cone

(D) Cube

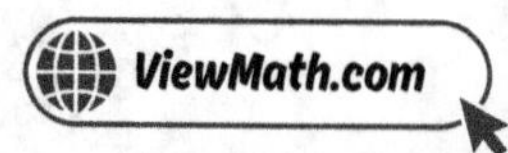

19. Rectangle $PQRS$ is similar to Rectangle $WXYZ$. $PQ = 5$, $QR = 8$, $WX = 12.5$. Find XY.

Your Answer:

20. The diagram shows a figure made of a rectangle and a triangle. What is the total area?

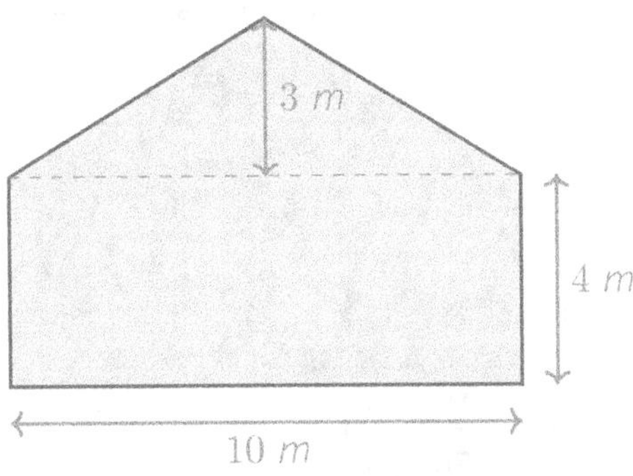

(A) $40\ m^2$

(B) $55\ m^2$

(C) $70\ m^2$

(D) $85\ m^2$

21. A shipping box is 12 in by 8 in by 6 in. What is the total surface area?

Your Answer:

22. Which formula is used to find the volume of any prism?

(A) $V = lwh$ only

(B) $V = Bh$

(C) $V = 2(lw + lh + wh)$

(D) $V = \frac{1}{3}Bh$

23. Which sampling method is most likely to produce a representative sample of a school's students?

(A) Survey students in the cafeteria during lunch

(B) Survey every fifth student on an alphabetical class list

(C) Survey students who volunteer to participate

(D) Survey only students in honors classes

 Find more at
ViewMath.com/TX-Grade7

 ViewMath.com

24. Three random samples of snack bar weights have means 49 g, 51 g, and 50 g. What is the best estimate for the population mean?

Your Answer:

25. Group X: mean 45, MAD 6. Group Y: mean 57, MAD 6. Express the difference in means as a multiple of MAD.

Your Answer:

26. How many total data values are in this stem-and-leaf plot?

Stem	Leaf
4	2 5 8
5	0 3 3 6 9
6	1 4

(A) 3

(B) 8

(C) 10

(D) 15

27. A survey of 60 people: 18 chose option A, 24 chose option B, 18 chose option C. Find the angle for option B in a circle graph.

Your Answer:

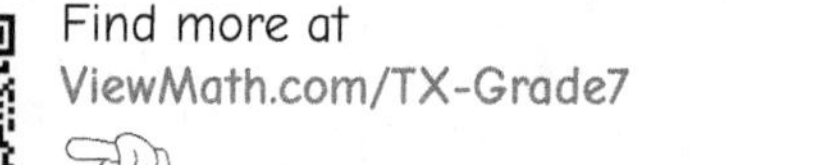
Find more at
ViewMath.com/TX-Grade7

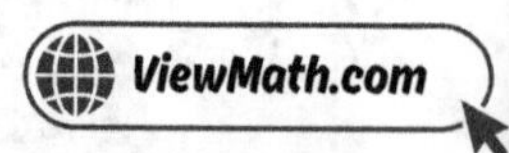

28. A bag contains 5 yellow and 5 purple marbles. Lisa picks a marble, records the color, and replaces it 40 times. She gets yellow 24 times. Which statement is correct?

(A) The model predicts $P(yellow) = 0.5$, and the experiment gives $P(yellow) = 0.6$, so there is a small difference

(B) The model is wrong because $24 \neq 20$

(C) The model predicts $P(yellow) = 0.6$

(D) The experiment proves yellow is more likely than purple

29. A coin is flipped and a die is rolled. What is the probability of getting tails and a 1? Write your answer as a fraction.

Your Answer:

30. How does increasing the number of trials in a simulation affect the results?

(A) The results become less reliable

(B) The experimental probability gets closer to the theoretical probability

(C) The results stay exactly the same

(D) The simulation takes less time

 End of Practice Test 8 

Great job finishing the test!

 My Score

I got ___________ out of 30 questions right.

Check your answers in the **Answer Key** at the back of the book.

Review any questions you missed. That's how we learn!

Check Your Score Online!

Visit **ViewMath Academy** to enter your answers and see which topics you need to review. You can also explore lessons, take quizzes, track your scores, and save your progress!

viewmath.com/score/7.1.TX.23

Or go to viewmath.com/score and enter code: 7.1.TX.23

Practice Test 9

 30 Questions

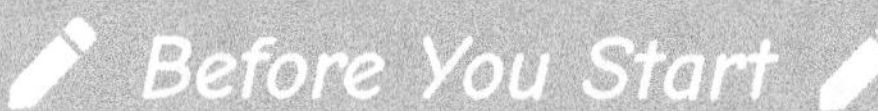

✏️ **Before You Start** ✏️

- ✔ **Read each question carefully** before choosing your answer.
- ✔ **Show your work** on scratch paper when you need to.
- ✔ **Skip hard questions** and come back to them later.
- ✔ **Check your answers** when you're done.
- ✔ **Take your time** — there's no rush!

⭐ You've Got This! ⭐

Do your best and show what you know!

1. Two students find k from the point $(8, 20)$. Student A says $k = 2.5$. Student B says $k = 0.4$. Who is correct?

(A) Student A, because $k = \frac{y}{x}$

(B) Student B, because $k = \frac{x}{y}$

(C) Both are correct, depending on which variable is independent

(D) Neither is correct

2. Five T-shirts cost $42.50. At this rate, how much do 8 T-shirts cost?

(A) $56.00

(B) $64.00

(C) $68.00

(D) $72.00

3. What is 40% of 150?

(A) 45

(B) 50

(C) 55

(D) 60

4. What multiplier represents a 40% decrease?

(A) 0.40

(B) 0.60

(C) 1.40

(D) 1.60

5. Maria earned $120 in simple interest on a $1,500 deposit at 4%. How long was the money deposited?

(A) 1 year

(B) 2 years

(C) 3 years

(D) 4 years

6. You deposit $800 in a savings account that pays 3% annual interest. How much interest do you earn in 1 year?

   (A) $3

   (B) $24

   (C) $80

   (D) $240

7. Income: $3,000. After all expenses, you save $450. What are your total expenses?

   (A) $3,450

   (B) $2,550

   (C) $2,450

   (D) $3,000

8. Complete the table for a $800 deposit at 5% compounded yearly. What is the balance at the end of Year 3?

Year	Interest Earned	Balance
0	—	$800
1	$40	$840
2	$42	$882
3	?	?

Your Answer

9. Which phrase best describes the expression $5n - 8$?

   (A) 8 less than 5 times a number

   (B) 5 less than 8 times a number

   (C) 8 minus 5 times a number

   (D) 5 times a number less 8 times that number

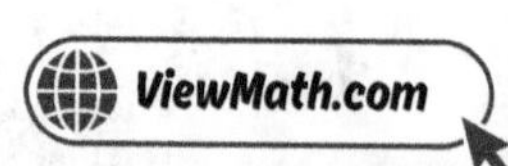

10. A rectangular garden has the dimensions shown. Write a simplified expression for the total distance around the garden (its perimeter).

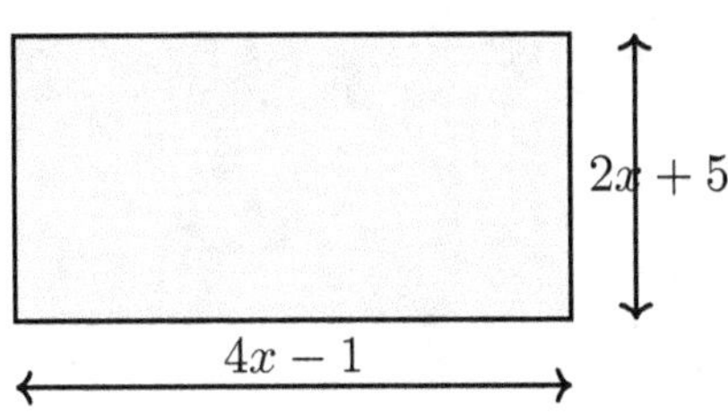

Your Answer:

11. Solve $5(n + 3) = -10$.

(A) $n = -5$

(B) $n = 1$

(C) $n = -1$

(D) $n = -\dfrac{25}{5}$

12. Solve $0.5x + 3 = 8$.

(A) $x = 10$

(B) $x = 22$

(C) $x = 5.5$

(D) $x = 2.5$

13. Which value of x is a solution to $4x - 1 > 11$?

(A) $x = 2$

(B) $x = 3$

(C) $x = 4$

(D) $x = 1$

14. Which graph represents $x \leq -2$?

(A) Open circle at -2, shade left

(B) Closed circle at -2, shade right

(C) Closed circle at -2, shade left

(D) Open circle at -2, shade right

Find more at
ViewMath.com/TX-Grade7

ViewMath.com

15. A model uses scale 1 : 25. A new model uses scale 1 : 75. A window that is 6 cm on the first model becomes how long on the second?

(A) 2 cm

(B) 6 cm

(C) 18 cm

(D) 150 cm

16. A square must have an area of 36 cm². How many different squares satisfy this condition?

(A) None

(B) Exactly one

(C) Exactly four

(D) More than one

17. Can a triangle have sides 4, 4, and 9?

(A) Yes, it is an isosceles triangle

(B) Yes, it is a scalene triangle

(C) No, the triangle inequality fails

(D) No, because two sides are equal

18. You slice a sphere with any flat plane. What shape is the cross-section?

(A) Oval

(B) Rectangle

(C) Circle

(D) It depends on the angle of the cut

19. Two similar triangles have corresponding sides of 7 and 21. What is the ratio of their areas?

(A) 1 : 3

(B) 1 : 7

(C) 1 : 9

(D) 7 : 21

Find more at
ViewMath.com/TX-Grade7

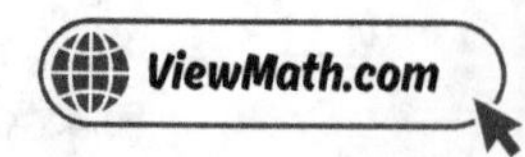

20. A shape is made of a rectangle 8 in by 6 in and two semicircles, each with a diameter of 6 in, attached to both short sides. What is the total area? Use $\pi \approx 3.14$.

(A) 48 in^2

(B) 76.26 in^2

(C) 104.52 in^2

(D) 133.04 in^2

21. A rectangular prism has a surface area of 94 cm^2. Its length is 5 cm and width is 3 cm. What is its height?

(A) 2 cm

(B) 3 cm

(C) 4 cm

(D) 5 cm

22. What is the volume of a rectangular prism with length 6 cm, width 4 cm, and height 3 cm?

(A) 13 cm^3

(B) 24 cm^3

(C) 48 cm^3

(D) 72 cm^3

23. Which question would require sampling rather than surveying the entire population?

(A) How many students are in a classroom of 25?

(B) What is the average lifespan of a light bulb produced by a factory?

(C) What is the total number of books on a shelf?

(D) How many chairs are in a room?

24. A teacher collects three random samples of 20 students from a school. The sample means for test scores are 78, 81, and 80. What is the best estimate for the population mean?

(A) 78

(B) 81

(C) 79.67

(D) 80

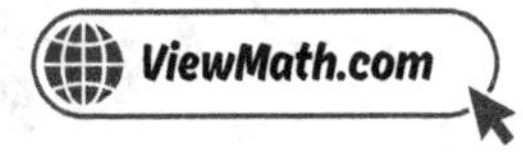

25. The table shows statistics for two basketball players' points per game over a season.

	Player A	Player B
Mean	18	20
Median	17	19
Range	10	25
MAD	3	7

Which player is more consistent?

(A) Player A, because the MAD and range are both smaller

(B) Player B, because the mean is higher

(C) Player A, because the mean is lower

(D) Player B, because the range is larger

26. A stem-and-leaf plot shows 12 data values. What is the median?

(A) The 6th value

(B) The 7th value

(C) The average of the 6th and 7th values

(D) The 12th value

27. The circle graph shows how a student spends a 24-hour day.

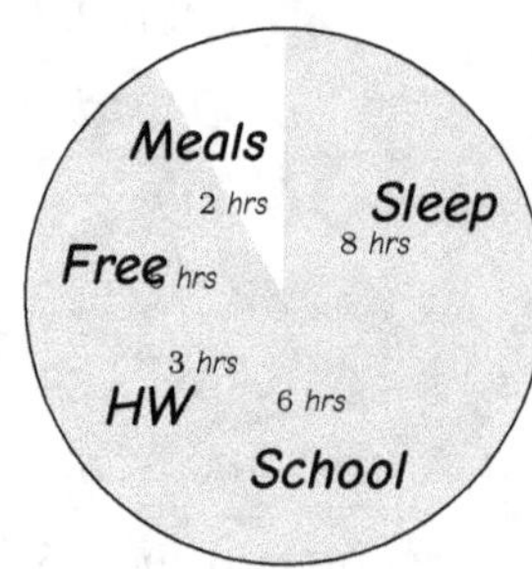

What percent of the day does the student spend on homework?

(A) 3%

(B) 8.3%

(C) 12.5%

(D) 25%

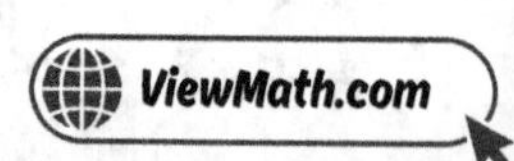

28. A teacher predicted that a spinner would land on red $\frac{1}{4}$ of the time. After 80 spins, red appeared 25 times. How do the observed and predicted results compare?

(A) They are very different — the model is wrong

(B) The observed ($\frac{25}{80} = 0.3125$) is close to the predicted (0.25)

(C) The observed result is exactly equal to the predicted result

(D) The predicted probability is higher than the observed

29. Three coins are flipped. What is the probability of getting NO heads (all tails)?

(A) $\frac{1}{2}$

(B) $\frac{1}{4}$

(C) $\frac{1}{8}$

(D) $\frac{3}{8}$

30. Describe how to use a standard die to simulate a $\frac{1}{3}$ probability event.

Your Answer:

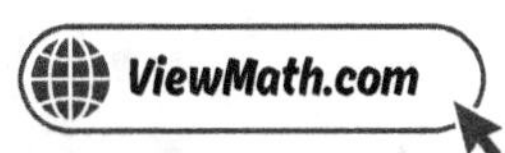

 End of Practice Test 9 

Great job finishing the test!

 My Score

I got ____________ out of 30 questions right.

Check your answers in the **Answer Key** at the back of the book.

💡 Review any questions you missed. That's how we learn!

📊 Check Your Score Online!

Visit **ViewMath Academy** to enter your answers and see which topics you need to review. You can also explore lessons, take quizzes, track your scores, and save your progress!

viewmath.com/score/7.1.TX.24

Or go to viewmath.com/score and enter code: 7.1.TX.24

Practice Test 10

30 Questions

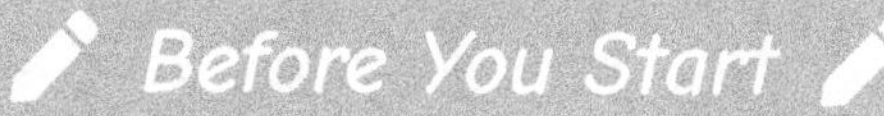

✏️ Before You Start ✏️

- ✔ **Read each question carefully** before choosing your answer.
- ✔ **Show your work** on scratch paper when you need to.
- ✔ **Skip hard questions** and come back to them later.
- ✔ **Check your answers** when you're done.
- ✔ **Take your time** — there's no rush!

★ You've Got This! ★

Do your best and show what you know!

1. The graph below shows a proportional relationship. What is the constant of proportionality?

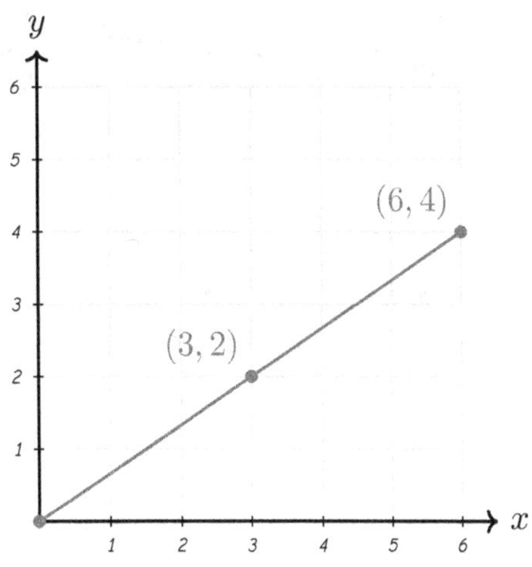

(A) $k = \frac{1}{2}$

(B) $k = \frac{2}{3}$

(C) $k = \frac{3}{2}$

(D) $k = 2$

2. The table shows the distance a bus travels over several hours. Using proportional reasoning, how far will the bus travel in 9 hours?

Hours	Miles
2	90
5	225
7	315
9	?

(A) 360 *miles*

(B) 385 *miles*

(C) 405 *miles*

(D) 450 *miles*

3. 18 *is what percent of 72?*

(A) 18%

(B) 20%

(C) 25%

(D) 36%

Find more at
ViewMath.com/TX-Grade7

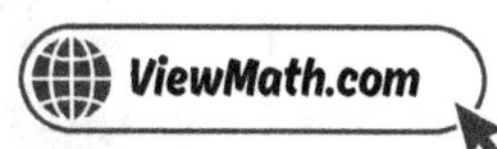

4. *Which of the following results in the greatest final value, starting from $100?*

   (A) *Increase by 30%*

   (B) *Increase by 20%, then increase by 10%*

   (C) *Increase by 15% twice*

   (D) *All give the same result*

5. *A $1,600 loan at 5.5% simple interest accrues $264 interest. How long was the loan?*

   (A) *2 years*

   (B) *3 years*

   (C) *4 years*

   (D) *5 years*

6. *The bar graph shows the annual fees at two banks and the annual interest earned on a $2,000 deposit. Which bank is the better deal, and how much more do you gain (or save) per year by choosing it?*

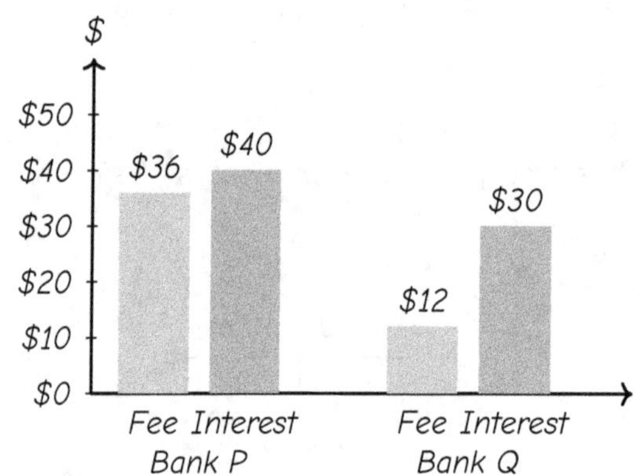

Your Answer:

7. *You want to save $1,800 in 12 months. How much must you save per month?*

Your Answer:

8. How much more does compound interest earn than simple interest on $500 at 8% for 3 years?

Your Answer

9. Evaluate $2a + 7$ when $a = -3$.

(A) 13   (B) 1

(C) -13 (D) -1

10. Simplify $3y + 8 - 5y - 2$.

(A) $8y + 6$ (B) $-2y + 6$

(C) $2y - 6$ (D) $-2y + 10$

11. The total area of the figure below is 84 square cm. Find x.

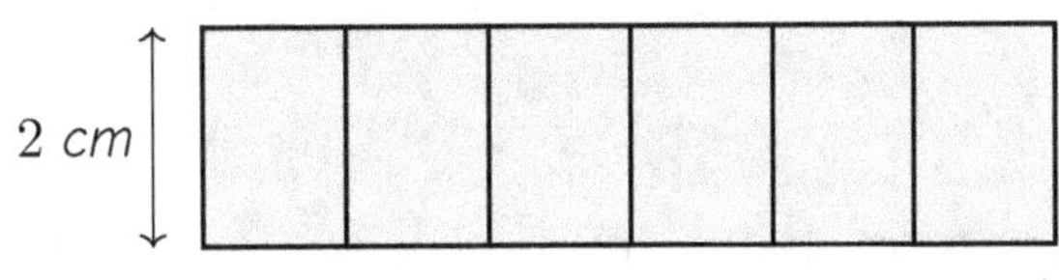

Your Answer

12. Solve $0.6x - 2.4 = 1.2$.

(A) $x = 6$ (B) $x = -2$

(C) $x = 2$ (D) $x = 0.6$

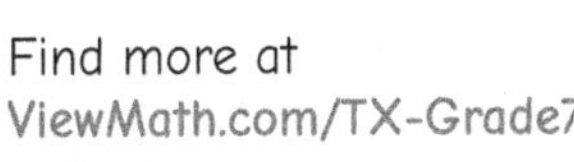
Find more at
ViewMath.com/TX-Grade7

13. *The balance scale tips to the left. Write and solve the inequality that describes this situation.*

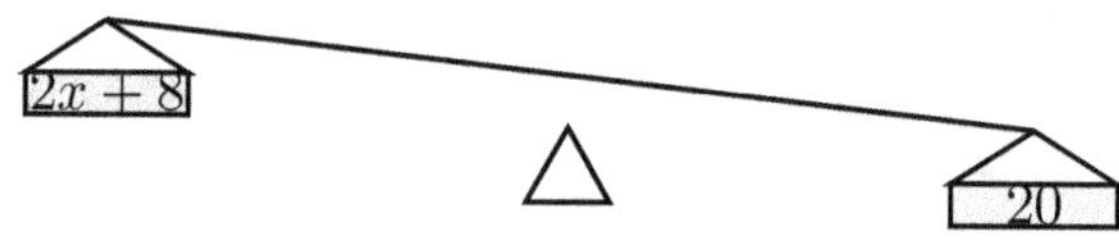

Your Answer

14. *Solve $6 - 2x \leq -4$ and describe the graph.*

(A) Closed circle at 5, shade right

(B) Open circle at 5, shade right

(C) Closed circle at 5, shade left

(D) Closed circle at −5, shade right

15. *A drawing uses 1 cm = 5 km. A road is 7 cm long. Another map shows the same road as 3.5 cm. What is the scale of the second map?*

(A) 1 cm = 2.5 km

(B) 1 cm = 5 km

(C) 1 cm = 10 km

(D) 1 cm = 17.5 km

16. *Can a triangle have sides 10 cm, 6 cm, and 3 cm? Explain why or why not.*

Your Answer

17. *Can a triangle have angles 90°, 100°, and −10°?*

(A) Yes, it is a right triangle

(B) Yes, because the sum is 180°

(C) No, because an angle cannot be negative

(D) No, because the sum is not 180°

Find more at
ViewMath.com/TX-Grade7

18. The diagram shows a cylinder with radius 4 cm and height 9 cm. A vertical cut is made through the center. Describe the cross-section and give its dimensions.

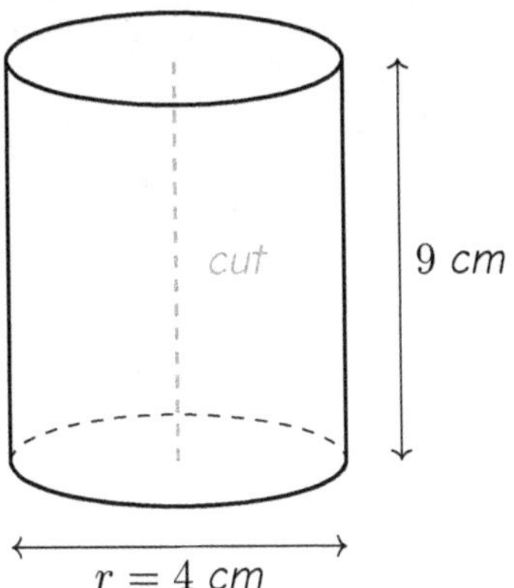

Your Answer:

19. A building is 60 ft tall and casts a 40-ft shadow. A nearby tree casts a 12-ft shadow. How tall is the tree?

(A) 8 ft

(B) 12 ft

(C) 18 ft

(D) 20 ft

20. A rectangular yard is 20 m by 15 m. A circular flower bed with radius 3 m is in the center. What is the area of the yard NOT covered by the flower bed? Use $\pi \approx 3.14$.

(A) 271.74 m^2

(B) 300 m^2

(C) 328.26 m^2

(D) 28.26 m^2

21. What is the surface area of a cube with side length 4 cm?

(A) 16 cm^2

(B) 64 cm^2

(C) 96 cm^2

(D) 24 cm^2

Find more at
ViewMath.com/TX-Grade7

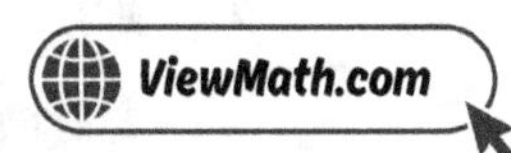

22. A triangular prism has a triangular base with legs 5 cm and 12 cm (right triangle). The prism is 15 cm long. What is the volume?

(A) 150 cm^3

(B) 300 cm^3

(C) 450 cm^3

(D) 900 cm^3

23. A radio station asks listeners to call in and vote for their favorite song. Explain why this is not a random sample.

Your Answer:

24. The table shows how 40 randomly selected students travel to school.

Transportation	Number
Bus	16
Car	10
Walk	8
Bike	6

The school has 600 students. Predict how many ride the bus.

(A) 16

(B) 160

(C) 200

(D) 240

25. Group A: $\{60, 65, 70, 75, 80\}$. Group B: $\{40, 55, 70, 85, 100\}$. Both have a mean of 70. Which group is more variable?

(A) Group A

(B) Group B

(C) They are equally variable

(D) Cannot be determined

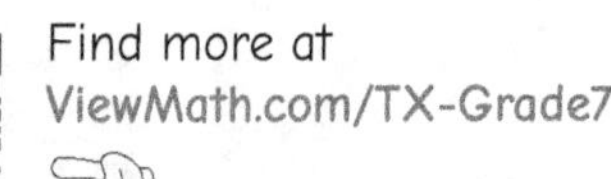
Find more at
ViewMath.com/TX-Grade7

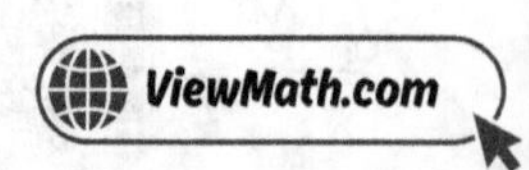

26. *The data set is:* $28, 31, 33, 36, 40, 42$. *What is the median?*

Your Answer:

27. *The full circle in a circle graph represents:*

(A) 50%

(B) 100% or 360°

(C) 180°

(D) The largest category

28. *Look at the spinner below. Which probability model correctly represents this spinner?*

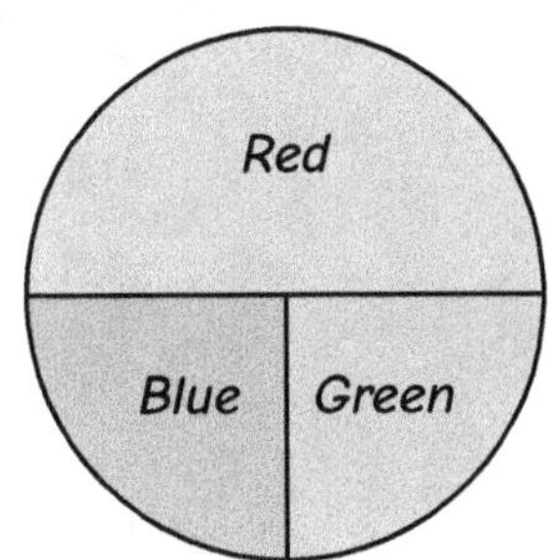

(A) $P(Red) = \frac{1}{3}$, $P(Blue) = \frac{1}{3}$, $P(Green) = \frac{1}{3}$

(B) $P(Red) = \frac{1}{2}$, $P(Blue) = \frac{1}{4}$, $P(Green) = \frac{1}{4}$

(C) $P(Red) = \frac{1}{4}$, $P(Blue) = \frac{1}{4}$, $P(Green) = \frac{1}{2}$

(D) $P(Red) = \frac{1}{2}$, $P(Blue) = \frac{1}{2}$, $P(Green) = 0$

29. *Two dice are rolled. What is the probability that the product (not the sum) of the two numbers is 12?*

(A) $\frac{2}{36}$

(B) $\frac{4}{36}$

(C) $\frac{3}{36}$

(D) $\frac{6}{36}$

30. A student simulates rolling two dice 36 times to see how often the sum is 7. They get a sum of 7 a total of 8 times. How does this compare to the expected number?

(A) Expected: 6; the simulation result (8) is slightly higher

(B) Expected: 6; the simulation result (8) is much higher

(C) Expected: 12; the simulation result (8) is lower

(D) Expected: 7; the simulation result (8) is exact

Find more at
ViewMath.com/TX-Grade7

⭐ End of Practice Test 10 ⭐

Great job finishing the test!

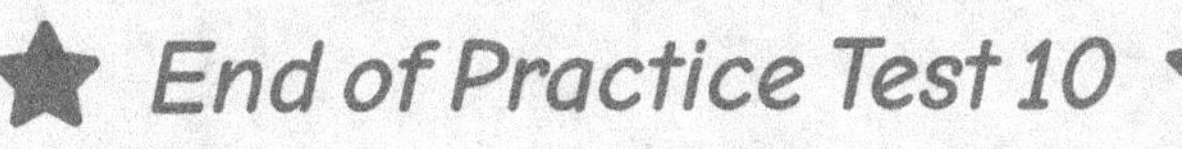 My Score

I got _____________ out of 30 questions right.

*Check your answers in the **Answer Key** at the back of the book.*

💡 *Review any questions you missed. That's how we learn!*

📊 Check Your Score Online!

Visit **ViewMath Academy** to enter your answers and see which topics you need to review. You can also explore lessons, take quizzes, track your scores, and save your progress!

viewmath.com/score/7.1.TX.25

Or go to viewmath.com/score and enter code: 7.1.TX.25

Answer Key & Explanations

Answer Key

First try each test on your own, then check your work here.

✅ Practice Test 1 — Answer Key

 $k = 2.5$; missing $x = 6$ 　　 C 　 D 　 D 　 C 　 $60 　 C 　 B

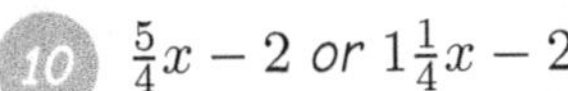 6 　 $\frac{5}{4}x - 2$ or $1\frac{1}{4}x - 2$ 　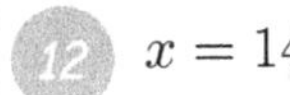 C 　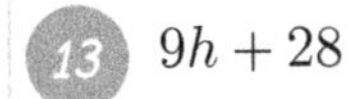 $x = 14$ 　 $9h + 28 > 100$; $h > 8$ 　 B

 5 cm 　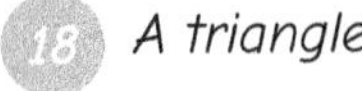 B 　 Yes 　 A triangle 　 C 　 B 　 A 　 B 　 B

 B 　 B 　 C 　 C 　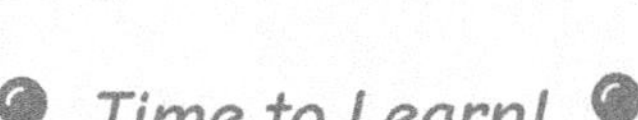 0.4 　 B 　 48

💡 Time to Learn! 💡

*Review the explanations below, **especially for the questions you missed**.*

Understanding why each answer is correct builds stronger problem-solving skills

Tip: *Circle any questions you got wrong, then read their explanation carefully.*

📖 Practice Test 1 — Detailed Explanations

1. $k = \frac{7.5}{3} = 2.5$. For the missing row: $15 = 2.5x$ so $x = 15 \div 2.5 = 6$.

2 Scale: 2 cm = 30 km, so 1 cm = 15 km. For 6.5 cm: $15 \times 6.5 = 97.5$ km.

3 The shaded portion is $3 \times 15 = 45$. The total is 75. Percent: $45 \div 75 = 0.60 = 60\%$.

4 Change: $14 - 8 = 6$. Percent increase: $6 \div 8 = 0.75 = 75\%$.

5 Rates over 2 years: \$500: $50/(500 \times 2) = 5\%$. \$1,000: $75/(1,000 \times 2) = 3.75\%$. \$800: $100/(800 \times 2) = 6.25\%$. The \$800 deposit had the highest rate.

6 $I = 500 \times 0.24 \times 0.5 = \$60$.

7 Groceries is a variable expense because the amount you spend on food changes month to month. Rent and phone plans are fixed amounts.

8 Year 1: $1,000 \times 1.06 = \$1,060$. Year 2: $1,060 \times 1.06 = \$1,123.60$.

9 Substitute: $\frac{5^2-1}{4} = \frac{25-1}{4} = \frac{24}{4} = 6$.

10 Find a common denominator: $\frac{2}{4}x + \frac{3}{4}x = \frac{5}{4}x$. The constant stays: $\frac{5}{4}x - 2$.

11 Distance from A to B is 15. $3(x+2) = 15$. Divide by 3: $x + 2 = 5$. Subtract 2: $x = 3$.

12 Subtract 1.5: $\frac{x}{4} = 3.5$. Multiply by 4: $x = 14$. The point is at 14 on the number line.

13 $9h + 28 > 100$. Subtract 28: $9h > 72$. Divide by 9: $h > 8$. You need to work more than 8 hours.

14 $x > 3$ uses an open circle because 3 is not included (strictly greater than).

Find more at
ViewMath.com/TX-Grade7

15 The scale ratio is $\frac{7}{7} = 1$. The length stays the same: $5 \times 1 = 5$ cm.

16 Two sides and the included angle (SAS) fully determine one unique triangle.

17 $11 + 13 = 24 > 20$ ✓, $11 + 20 = 31 > 13$ ✓, $13 + 20 = 33 > 11$ ✓. All three checks pass.

18 A vertical slice through the apex of a cone produces an isosceles triangle. The base of the triangle is a diameter of the cone's base.

19 All circles have the same shape. Any circle can be scaled (dilated) to match any other circle. Therefore, all circles are similar to each other.

20 Rectangle: $6 \times 4 = 24$ cm^2. Triangle: $\frac{1}{2} \times 6 \times 2 = 6$ cm^2. Total: $24 + 6 = 30$ cm^2.

21 Two triangles: $2 \times \frac{1}{2} \times 6 \times 4 = 24$ cm^2. Three rectangles: $6 \times 10 + 5 \times 10 + 5 \times 10 = 60 + 50 + 50 = 160$ cm^2. Total: $24 + 160 = 184$ cm^2.

22 $s^3 = 64$. $s = 4$ cm because $4^3 = 64$.

23 Only surveying recent buyers introduces bias — they may feel differently than all customers.

24 $10{,}000 \times 200 = 2{,}000{,}000$ grams.

25 $|4 - 8| + |6 - 8| + |8 - 8| + |10 - 8| + |12 - 8| = 4 + 2 + 0 + 2 + 4 = 12$. MAD $= \frac{12}{5} = 2.4$.

26 Values greater than 25: $27, 31, 35, 39, 42$ — that's 5 values.

27 $\frac{16}{40} = 0.40 = 40\%$.

Find more at
ViewMath.com/TX-Grade7

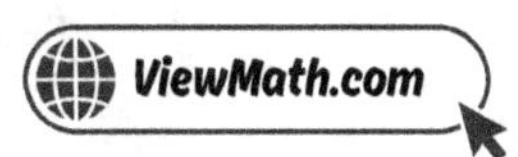

28 $P(Z) = 1 - 0.15 - 0.45 = 0.40.$

29 $P(\text{spinner} = 3) = \frac{1}{4}, P(\text{die} = 3) = \frac{1}{6}. P = \frac{1}{4} \times \frac{1}{6} = \frac{1}{24}.$

30 $80 \times 0.60 = 48.$

✔ Practice Test 2 — Answer Key

1 B	2 21 feet	3 90 students	4 C	5 A
6 B	7 C	8 $1,685.40		

9 C	10 $5a + 2b$	11 B	12 $x = 8$	13 $9s + 35 \leq 200; s \leq 18$
14 $x \geq -6$	15 B			

16 D	17 B	18 B	19 12.8 ft	20 B
21 C	22 $9\ ft^3$	23 C	24 900	

25 A	26 Range = 67; Median = 33.5	27 B	28 B
29 C	30 B		

💡 Time to Learn! 💡

Review the explanations below, **especially for the questions you missed**.

Understanding why each answer is correct builds stronger problem-solving skills.

Tip: Circle any questions you got wrong, then read their explanation carefully.

📖 Practice Test 2 — Detailed Explanations

1 $k = \frac{54}{2} = 27, \frac{135}{5} = 27, \frac{216}{8} = 27.$ The car gets 27 miles per gallon.

Find more at
ViewMath.com/TX-Grade7

2   *Scale: 1 inch = 6 feet. For 3.5 inches: $6 \times 3.5 = 21$ feet.*

3 45% *of 200: $0.45 \times 200 = 90$ students take the bus.*

4   *A 25% increase means you keep 100% and add 25%: $1 + 0.25 = 1.25$.*

5   *6 months $= 0.5$ years. $I = 900 \times 0.08 \times 0.5 = \$36$.*

6   *Interest $= 250 \times 0.20 = \$50$. Total owed $= 250 + 50 = \$300$.*

7   *Rent stays the same each month, making it a fixed expense. The others vary.*

8   *$A = 1{,}500(1.06)^2 = 1{,}500 \times 1.1236 = \$1{,}685.40$.*

9   *Substitute: $\frac{10}{2} - 4 = 5 - 4 = 1$.*

10   *Combine a-terms: $-2a + 7a = 5a$. Combine b-terms: $5b - 3b = 2b$. Result: $5a + 2b$.*

11   *Each pays $\frac{c}{3} + 4 = 16$. This means $\frac{c}{3} = 12$, so $c = 36$. Or: total paid $= 3 \times 16 = 48$, wrapping $= 3 \times 4 = 12$, gift $= 48 - 12 = 36$.*

12   *Subtract 4.2: $-0.3x = -2.4$. Divide by -0.3: $x = 8$. Check: $-0.3(8) + 4.2 = -2.4 + 4.2 = 1.8$ ✓.*

13   *Total cost: $9s + 35 \leq 200$. Subtract 35: $9s \leq 165$. Divide by 9: $s \leq 18.\overline{3}$. Since s must be a whole number, $s \leq 18$.*

14   *Closed circle means -6 is included, shading right means values greater than or equal to -6: $x \geq -6$.*

15. A smaller scale means each cm represents more real distance, so fewer cm are needed. The drawing gets smaller. The shape stays the same.

16. Three angles (AAA) determine the shape but not the size. You can draw infinitely many similar triangles of different sizes with these angles.

17. $180° - 35° - 75° = 70°$.

18. A diagonal plane that passes through all six faces of a cube creates a regular hexagonal cross-section.

19. $\frac{4}{5} = \frac{h}{16}$. Cross-multiply: $5h = 64$. Divide: $h = 12.8$ ft.

20. Square: $8^2 = 64$ cm^2. Semicircle: $\frac{1}{2} \times \pi r^2 = \frac{1}{2} \times 3.14 \times 16 = 25.12$ cm^2. Total: $64 + 25.12 = 89.12$ cm^2.

21. $SA = 2(3 \times 2) + 2(3 \times 1) + 2(2 \times 1) = 12 + 6 + 4 = 22$ m^2. Cans: $22 \div 5 = 4.4$. Round up to 5 cans.

22. $V = 3 \times 1.5 \times 2 = 9$ ft^3.

23. In a random sample, every member of the population has an equal chance of being chosen.

24. $\frac{3}{8} = 37.5\%$. Then 37.5% of $2,400 = 900$.

25. Class M: $Q_3 - Q_1 = 17 - 12 = 5$. Class N: $18 - 13 = 5$. Both have IQR $= 5$.

26. Data (20 values): $5, 8, 9, 12, 14, 16, 20, 25, 28, 32, 35, 37, 39, 41, 45, 50, 58, 63, 67, 72$. Range $= 72 - 5 = 67$. Median $= \frac{32+35}{2} = 33.5$.

27. A circle graph (pie chart) shows how a whole quantity is divided into categories.

Find more at
ViewMath.com/TX-Grade7

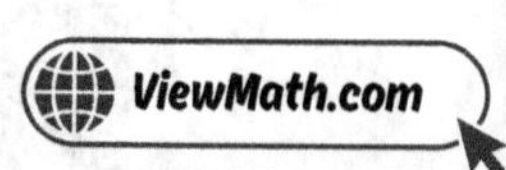

28. Each color should appear about 50 times (200×0.25). The results $(55, 48, 52, 45)$ are all close to 50, supporting the model.

29. Outcomes with at least one tail: HT, TH, TT — that is 3 out of 4. $P = \frac{3}{4}$.

30. With 4 equal sections, each should appear about 25 times in 100 spins. The results $(26, 24, 28, 22)$ are all close to 25, suggesting the spinner is fair.

✅ Practice Test 3 — Answer Key

1. B	2. C	3. C	4. B	5. C
6. C	7. A	8. B	9. B	10. A

11. B   12. $x = 7$ 13. B   14. $x \leq -3$; closed circle at -3, shade left   15. A   16. C

17. B   18. B   19. B   20. C   21. C   22. 126 cm^3 23. B   24. B   25. B

26. 11   27. A   28. B   29. C   30. A

💡 Time to Learn! 💡

Review the explanations below, **especially for the questions you missed**.

Understanding why each answer is correct builds stronger problem-solving skills.

Tip: Circle any questions you got wrong, then read their explanation carefully.

📖 Practice Test 3 — Detailed Explanations

1. If $k = \frac{3}{4}$, then $y = \frac{3}{4}x$. For $x = 8$: $y = \frac{3}{4} \times 8 = 6$. So $(8, 6)$ works.

Find more at
ViewMath.com/TX-Grade7

(2) Per person: $\frac{3}{2} \div 4 = \frac{3}{8}$ cup. For 10: $\frac{3}{8} \times 10 = \frac{30}{8} = 3\frac{3}{4}$ cups.

(3) Divide: $63 \div 0.90 = 70$.

(4) Change: $13{,}200 - 12{,}000 = 1{,}200$. Percent increase: $1{,}200 \div 12{,}000 = 0.10 = 10\%$.

(5) $I = 1{,}200 \times 0.035 \times 2 = \$84$.

(6) Interest $= 300 \times 0.18 \times 1 = \$54$.

(7) Budget for 5 months: $80 \times 5 = \$400$. Spent: $\$350$. Under by $400 - 350 = \$50$.

(8) Year 1: $500 \times 1.10 = \$550$. Year 2: $550 \times 1.10 = \$605$.

(9) Substitute: $(-2)^2 + 3(-2) = 4 + (-6) = -2$.

(10) $-6y + 4y = -2y$. Then $-2y - y = -3y$.

(11) The student wrote $3x + 4$ instead of $3x + 12$. Distributing correctly gives $3x + 12 = 30$.

(12) Add 7.5: $2.5x = 17.5$. Divide by 2.5: $x = 7$. Check: $2.5(7) - 7.5 = 17.5 - 7.5 = 10$ ✓.

(13) Divide by -4 and flip the sign: $x < -5$.

(14) Subtract 15: $-5x \geq 15$. Divide by -5 and flip: $x \leq -3$. Closed circle at -3, shade left.

(15) Scale ratio: $\frac{10}{40} = \frac{1}{4}$. Every old length is multiplied by $\frac{1}{4}$ on the new drawing.

Find more at
ViewMath.com/TX-Grade7

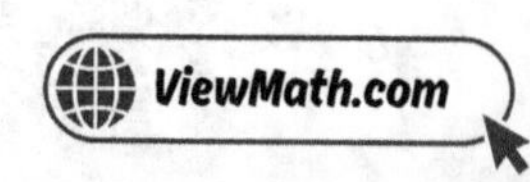

16 Check: $9^2 + 12^2 = 81 + 144 = 225 = 15^2$. Since the Pythagorean theorem holds, this is a right triangle.

17 SSS with valid side lengths ($3 + 4 = 7 > 5$, etc.) produces exactly one unique triangle. AAA gives many triangles, and one side alone is insufficient.

18 A vertical cut through the apex of a cone produces a triangle (specifically an isosceles triangle).

19 $\frac{5}{3} = \frac{h}{15}$. Cross-multiply: $3h = 75$. Divide: $h = 25$ ft.

20 Horizontal: $12 \times 2 = 24$ cm^2. Vertical: $2 \times 8 = 16$ cm^2. Total: $24 + 16 = 40$ cm^2.

21 Surface area depends on the square of the dimensions. Doubling all dimensions multiplies each face area by $2^2 = 4$, so the total surface area quadruples.

22 Bottom prism: $10 \times 3 \times 3 = 90$ cm^3. Top prism: $4 \times 3 \times 3 = 36$ cm^3. Total: $90 + 36 = 126$ cm^3.

23 A random sample provides an estimate. About 35% of the population likely prefers pizza, but the exact percentage may differ slightly.

24 Larger random samples tend to give more reliable estimates because they better represent the population.

25 A wider range indicates more variability, not better performance. Class A has a higher median and is more consistent (smaller IQR and range).

26 Count leaves: $3 + 5 + 2 + 1 = 11$.

27 Class A: 45% of $20 = 9$. Class B: 30% of $30 = 9$. Same number: 9 each.

28 The probability of drawing red ($\frac{4}{5}$) is different from drawing blue ($\frac{1}{5}$), so this is a non-uniform model.

Find more at
ViewMath.com/TX-Grade7

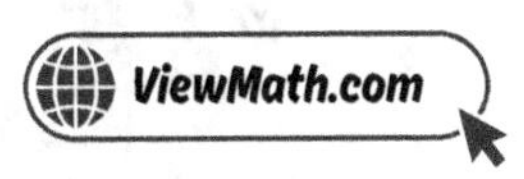

29) The only way to get a sum of 2 is $(1,1)$. $P = \frac{1}{36}$.

30) $P = \frac{6}{40} = 0.15$. The theoretical probability is $\frac{1}{8} = 0.125$, close to the simulation result.

✅ Practice Test 4 — Answer Key

1) B	2) 9 trips	3) D
4) 15%	5) C	6) B
7) A	8) \$1,197.90	9) C

10) A 11) C 12) $x = 12$ 13) B 14) $x > 2$; open circle at 2, shade right 15) C 16) B

17) C 18) C 19) A 20) B 21) C 22) B 23) B 24) 200 25) A 26) A

27) B 28) C 29) $\frac{3}{8}$ 30) B

💡 Time to Learn! 💡

Review the explanations below, **especially for the questions you missed**.

Understanding why each answer is correct builds stronger problem-solving skills.

Tip: Circle any questions you got wrong, then read their explanation carefully.

📖 Practice Test 4 — Detailed Explanations

1) $k = \frac{y}{x} = \frac{12}{3} = 4$. Check: $\frac{20}{5} = 4$ and $\frac{32}{8} = 4$.

2) Per trip: $540 \div 3 = 180$ packages. Trips needed: $1,620 \div 180 = 9$.

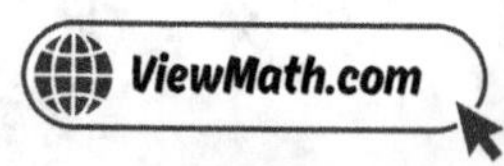

3   Divide the part by the percent: $9 \div 0.12 = 75$.

4   Change: $400 - 340 = 60$. Percent decrease: $60 \div 400 = 0.15 = 15\%$.

5 $I = Prt = 500 \times 0.06 \times 3 = \$90$.

6   A credit union works like a bank (savings, checking, loans) but is owned by its members and often has lower fees.

7   Income − Expenses = Savings, which can be rearranged to Income = Savings + Expenses.

8   Year 1: $900 \times 1.10 = 990$. Year 2: $990 \times 1.10 = 1{,}089$. Year 3: $1{,}089 \times 1.10 = \$1{,}197.90$.

9   "The product of 7 and p" is $7p$. "Decreased by 12" means subtract 12: $7p - 12$.

10 $7x$ and $3x$ are like terms. Add the coefficients: $7 + 3 = 10$. Result: $10x$.

11   Divide by 7: $x - 2 = 0$. Add 2: $x = 2$. Check: $7(2 - 2) = 7(0) = 0$ ✓.

12   Multiply every term by 4: $2x + x = 36$. Combine: $3x = 36$. Divide by 3: $x = 12$. Check: $\frac{12}{2} + \frac{12}{4} = 6 + 3 = 9$ ✓.

13   The graph shows $x \leq -2$ (closed circle, shading left). $-4x \geq 8$: divide by -4 and flip: $x \leq -2$ ✓.

14   Subtract 10: $-4x < -8$. Divide by -4 and flip: $x > 2$. Open circle at 2, shade right.

15   Scale ratio: $\frac{20}{10} = 2$. Every length on the new drawing is 2 times the old one. The drawing gets bigger because each cm now represents fewer real km.

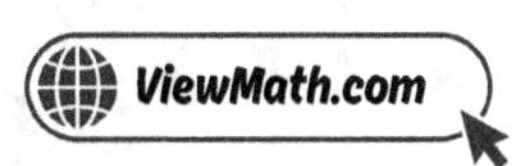

(16) The four rectangles all have perimeter 16 cm $(2(7+1) = 2(6+2) = 2(5+3) = 2(4+4) = 16)$ but different shapes. A given perimeter allows many rectangles.

(17) $5^2 + 12^2 = 25 + 144 = 169 = 13^2$. The Pythagorean theorem holds, so it is a right triangle.

(18) Any cut parallel to the base of a prism gives a shape congruent to the base. A hexagonal prism has a hexagonal base, so the cross-section is a hexagon.

(19) Scale factor: $\frac{5}{10} = 0.5$. $YZ = 14 \times 0.5 = 7$.

(20) Rectangle: $18 \times 6 = 108$ in^2. Triangle: $\frac{1}{2} \times 6 \times 4 = 12$ in^2. Banner: $108 - 12 = 96$ in^2.

(21) A rectangular prism has 6 faces: top, bottom, front, back, left, and right.

(22) Volume measures three-dimensional space and is measured in cubic units.

(23) Selecting every 10th student from the school roster is systematic and gives all students a chance, making it the least biased.

(24) $\frac{10}{40} = 25\%$. Then 25% of $800 = 200$.

(25) Difference $= 60 - 50 = 10$. In MADs: $\frac{10}{3} \approx 3.3$.

(26) Stems are tens digits $(4, 5, 6)$, leaves are ones digits listed in order. Only choice A has correct stems and ordered leaves.

(27) $100 + 80 + 60 + 100 = 340°$, which is 20° short of 360°. The graph is incorrect.

(28) $P(tails) = 1 - 0.60 = 0.40$. Since heads and tails have different probabilities, this is a non-uniform model.

29 Outcomes with exactly 2 tails: HTT, THT, TTH — that is 3 out of 8. $P = \frac{3}{8}$.

30 Each coin flip has a 50% chance of heads (pass). Flipping 3 coins and counting exactly 2 heads simulates exactly 2 passes.

✅ Practice Test 5 — Answer Key

1 $k = 70$ widgets per hour   **2** C   **3** B   **4** A   **5** C

6 Advantage: buy now, pay later; Disadvantage: you may pay interest   **7** C   **8** B   **9** B

10 $10x + 5$ **11** $x = 4$ **12** B   **13** $x \le -5$ **14** C   **15** C   **16** 4   **17** C   **18** A

19 B   **20** A   **21** 8 in   **22** C   **23** B   **24** B   **25** A   **26** A   **27** 54°   **28** C

29 $\frac{1}{9}$ **30** C

💡 Time to Learn! 💡

Review the explanations below, **especially for the questions you missed**.

Understanding why each answer is correct builds stronger problem-solving skills.

Tip: Circle any questions you got wrong, then read their explanation carefully.

📖 Practice Test 5 — Detailed Explanations

1 $k = \frac{175}{2.5} = 70$ widgets per hour.

Find more at
ViewMath.com/TX-Grade7

2 Set up a proportion: $\frac{2}{15} = \frac{7}{d}$. Cross-multiply: $2d = 105$, so $d = 52.5$ miles.

3 Divide: $54 \div 200 = 0.27 = 27\%$.

4 April to May: $(500 - 400) \div 400 = 0.25 = 25\%$. May to June: $(550 - 500) \div 500 = 0.10 = 10\%$. The greatest percent change is 25%.

5 $300 = 2{,}000 \times 0.03 \times t$ ⮕ $300 = 60t$ ⮕ $t = 5$ years.

6 Credit cards let you purchase before having the cash, but you owe interest if you don't pay the full balance on time.

7 $\frac{240}{2{,}400} = 0.10 = 10\%$.

8 $48 = 800 \times r \times 2$. So $r = \frac{48}{1{,}600} = 0.03 = 3\%$.

9 "The sum of k and 10" is $k + 10$. "Half" of that sum is $\frac{k+10}{2}$.

10 Combine variable terms: $3x + 5x + 2x = 10x$. Combine constants: $2 - 1 + 4 = 5$. Perimeter: $10x + 5$.

11 Divide by 4: $x + 9 = 13$. Subtract 9: $x = 4$. Check: $4(4 + 9) = 4(13) = 52$ ✓.

12 Multiplying every term by 4 (the LCD) clears the fractions: $x + 2 = 12$.

13 Subtract 8: $-3x \geq 15$. Divide by -3 and flip: $x \leq -5$.

14 $x \geq -4$ includes -4 and everything to the right. $-5 < -4$, so -5 is not a solution.

Find more at
ViewMath.com/TX-Grade7

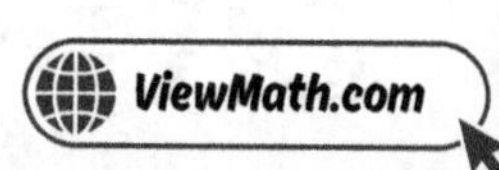

15   Scale ratio: $\frac{5}{2} = 2.5$. New dimensions: $3 \times 2.5 = 7.5$ cm and $4 \times 2.5 = 10$ cm. New area: $7.5 \times 10 = 75$ cm^2.

16   Factor pairs of 24: 1×24, 2×12, 3×8, 4×6. That is 4 different rectangles.

17   AAA alone does not fix the size of the triangle. Many triangles can share these angle measures.

18   A horizontal cut through a cone parallel to the base produces a circle smaller than the base. The higher you cut, the smaller the circle.

19   Scale factor: $\frac{20}{5} = 4$. Shortest side: $3 \times 4 = 12$.

20   Room: $14 \times 10 = 140$ ft^2. Closet: $3 \times 4 = 12$ ft^2. Remaining: $140 - 12 = 128$ ft^2.

21 $s^2 = 384 \div 6 = 64$. So $s = 8$ in.

22 $V = 50 \times 30 \times 40 = 60{,}000$ cm^3.

23   The sample size is the number of individuals selected — 500 patients.

24   The sample mean is a good estimate of the population mean, but may not be exact due to random variation.

25   Difference $= 82 - 76 = 6$. In MADs: $\frac{6}{5} = 1.2$. Since $1.2 < 2$, the difference is not considered very meaningful.

26   The values are $22, 22, 25, 28$. The value 22 appears twice (most often), so it is the mode.

27 $0.15 \times 360° = 54°$.

28   There are 2 B tiles out of 5 total. $P(B) = \frac{2}{5}$.

Find more at
ViewMath.com/TX-Grade7

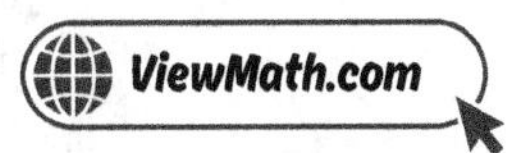

29 Products of 6: $(1,6), (2,3), (3,2), (6,1)$ — 4 outcomes. $P = \frac{4}{36} = \frac{1}{9}$.

30 Before choosing a model or running trials, you must first identify what event you're simulating and what the possible outcomes are.

✓ Practice Test 6 — Answer Key

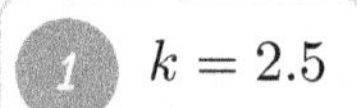 $k = 2.5$  C  C  B  C  B  B  B  C

 A  A 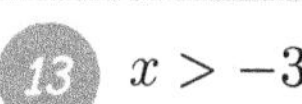 A  $x > -3$  A  C  B  No  No

 B  50 m^2  C  240 m^3  C  C  B  C 27 D

28 B 29 C 30 B

💡 Time to Learn! 💡

Review the explanations below, **especially for the questions you missed**.

Understanding why each answer is correct builds stronger problem-solving skills.

Tip: Circle any questions you got wrong, then read their explanation carefully.

📖 Practice Test 6 — Detailed Explanations

1 $k = \frac{y}{x} = \frac{15}{6} = 2.5$.

2 Rate: $120 \div 8 = 15$ pages/min. Time for 450: $450 \div 15 = 30$ min.

Find more at
ViewMath.com/TX-Grade7

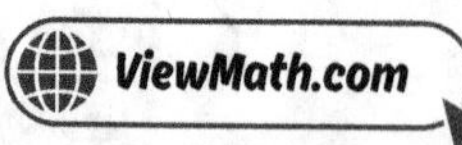

3. $35\% = 0.35$. *Multiply:* $0.35 \times 480 = 168$ *fiction books.*

4. *Multiply by the decrease multiplier:* $200 \times 0.65 = \$130$.

5. *A:* $1{,}000 \times 0.05 \times 2 = \$100$. *B:* $1{,}000 \times 0.02 \times 5 = \$100$. *Both earn* $\$100$.

6. *Interest* $= 500 \times 0.12 \times 1 = \$60$. *Total* $= 500 + 60 = \$560$.

7. *Interest* $= 500 \times 0.02 = \$10$.

8. *Year 1:* $700 \times 1.05 = \$735$. *Year 2:* $735 \times 1.05 = \$771.75$.

9. *"Divided by 4" gives* $\frac{n}{4}$. *"Increased by 3" means add 3:* $\frac{n}{4} + 3$.

10. *Combine variable terms:* $-4x + 7x = 3x$. *Combine constants:* $9 - 3 = 6$. *Result:* $3x + 6$.

11. *Divide by 10:* $x - 5 = 3$. *Add 5:* $x = 8$. *Check:* $10(8 - 5) = 10(3) = 30$ ✓.

12. *From* $x = 2$ *to* $x = 4$, *y increases by 1 while x increases by 2, so rate* $= 0.5$. *When* $x = 2$: $0.5(2) + 1.5 = 2.5$ ✓.

13. *Add 4:* $-7x < 21$. *Divide by* -7 *and flip:* $x > -3$.

14. *You can spend at most* $\$30$: $5h \le 30$, *so* $h \le 6$. *Closed circle at 6, shade left (and* $h \ge 0$).

15. *Scale ratio:* $\frac{12}{4} = 3$. *New width:* $0.5 \times 3 = 1.5$ *in.*

Find more at
ViewMath.com/TX-Grade7

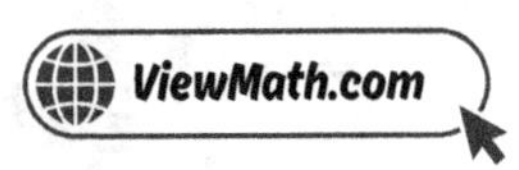

16. All circles with radius 5 cm are congruent — they have the same shape and size. In terms of the figure itself (ignoring position), there is exactly one such circle.

17. $8 + 8 = 16$, which is not greater than 16. The triangle inequality requires the sum to be strictly greater.

18. A rectangular prism has only flat faces and straight edges. All cross-sections are polygons (triangles, rectangles, pentagons, hexagons). A circular cross-section requires a curved surface.

19. Smaller side $= 30 \times \frac{2}{5} = 12$ cm.

20. $A = \frac{1}{2}(b_1 + b_2) \times h = \frac{1}{2}(8 + 12) \times 5 = \frac{1}{2} \times 20 \times 5 = 50$ m^2.

21. $SA = 2(6 \times 6) + 2(6 \times 10) + 2(6 \times 10) = 72 + 120 + 120 = 312$ ft^2.

22. $B = \frac{1}{2} \times 10 \times 6 = 30$ m^2. $V = 30 \times 8 = 240$ m^3.

23. When people choose whether to participate, it is a voluntary response sample, which tends to be biased.

24. Using multiple random samples and averaging reduces the effect of random variation.

25. Same mean means same center. Different MADs mean one group's data is more spread out than the other's.

26. A stem-and-leaf plot shows every individual data value AND the overall shape/distribution of the data.

27. 55% of $500 = 0.55 \times 500 = 275$.

28. The probabilities are $0.4, 0.3, 0.2, 0.1$ — all different, so this is non-uniform. They sum to 1.0, making it valid.

Find more at
ViewMath.com/TX-Grade7

29. $P(\text{tails}) = \frac{1}{2}$ and $P(\text{even}) = \frac{3}{6} = \frac{1}{2}$. $P = \frac{1}{2} \times \frac{1}{2} = \frac{1}{4}$. Or: 3 favorable out of 12 total outcomes.

30. A simulation uses random numbers, coins, dice, or spinners to model a real-world event and estimate probabilities.

☑ Practice Test 7 — Answer Key

1. $y = 36$
2. C
3. 25%
4. B
5. 3 years
6. C
7. $480
8. B
9. B
10. A
11. $x = 7$
12. C
13. A
14. C
15. 15 cm
16. B
17. B
18. B
19. B
20. 52 cm^2
21. C
22. C
23. 500
24. B
25. IQR is not affected by outliers
26. B
27. Sports: 40%, 144°; Music: 25%, 90°; Art: 15%, 54°; Reading: 20%, 72°
28. B
29. B
30. C

💡 Time to Learn! 💡

Review the explanations below, **especially for the questions you missed**.

Understanding why each answer is correct builds stronger problem-solving skills

Tip: Circle any questions you got wrong, then read their explanation carefully.

📖 Practice Test 7 — Detailed Explanations

1. $k = \frac{20}{5} = 4$. So $y = 4 \times 9 = 36$.

Find more at
ViewMath.com/TX-Grade7

2. *Rate:* $350 \div 5 = 70$ *bottles/hr. In 12 hours:* $70 \times 12 = 840$ *bottles.*

3. *Divide the part by the whole:* $42 \div 168 = 0.25 = 25\%$.

4. *Change:* $75 - 60 = 15$. *Percent decrease:* $15 \div 75 = 0.20 = 20\%$.

5. $450 = 5{,}000 \times 0.03 \times t \; \boxtimes \; 450 = 150t \; \boxtimes \; t = 3$ *years.*

6. *A debit card takes money directly from your bank account, so you must already have the funds.*

7. $3{,}200 \times 0.15 = \$480$.

8. *Compound interest grows faster because you earn interest on your interest, making the base larger each year.*

9. *"Twice a number" is* $2n$. *"6 more than" means add 6:* $2n + 6$.

10. *Combine:* $\frac{3}{4}n - \frac{1}{4}n = \frac{2}{4}n = \frac{1}{2}n$. *Add the constant:* $\frac{1}{2}n + 5$.

11. $3(2x + 5) = 57$. *Divide by 3:* $2x + 5 = 19$. *Subtract 5:* $2x = 14$. *Divide by 2:* $x = 7$.

12. *Subtract 4.5:* $-1.5x = -7.5$. *Divide by* -1.5: $x = 5$. *Check:* $-1.5(5) + 4.5 = -7.5 + 4.5 = -3$ ✓.

13. *Subtract 4:* $\frac{x}{3} < 6$. *Multiply by 3:* $x < 18$.

14. *Open circle at 0 (not included) with shading left means* $x < 0$, *which is all numbers less than 0.*

15. *Scale ratio:* $\frac{12}{4} = 3$. *New length:* $5 \times 3 = 15$ *cm.*

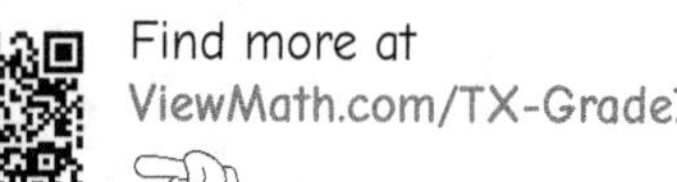

Find more at
ViewMath.com/TX-Grade7

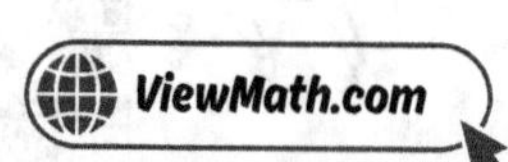

16. All three inequality checks pass $(5 + 5 = 10 > 5)$. Three equal sides (SSS) produce exactly one equilateral triangle.

17. SSS with valid lengths produces exactly one unique triangle. Check: $6 + 8 = 14 > 10$, $6 + 10 = 16 > 8$, $8 + 10 = 18 > 6$. All pass.

18. The shape of a cross-section depends on where you cut and at what angle. Different cuts through the same figure can produce different shapes.

19. Scale factor: $\frac{15}{6} = 2.5$. Shorter side: $4 \times 2.5 = 10$ in.

20. Top: $10 \times 2 = 20$ cm^2. Bottom: $10 \times 2 = 20$ cm^2. Middle: $2 \times 6 = 12$ cm^2. Total: $20 + 20 + 12 = 52$ cm^2.

21. $SA = 2(30 \times 20) + 2(30 \times 6) + 2(20 \times 6) = 1200 + 360 + 240 = 1,800$ cm^2.

22. $V = 2 \times 1.5 \times 3 = 9$ ft^3.

23. $\frac{20}{80} = 25\%$. Then 25% of $2,000 = 500$.

24. Average: $\frac{17+20+15}{3} \approx 17.33$ out of $50 \approx 34.67\%$. Then 34.67% of $800 \approx 277$.

25. Range uses only the extreme values, which can be misleading with outliers. IQR measures the middle 50% and gives a more stable picture of typical variability.

26. The key tells readers what the stems and leaves represent, so they can correctly read the data values.

27. Sports: $\frac{48}{120} = 40\%$, $0.40 \times 360° = 144°$. Music: $\frac{30}{120} = 25\%$, $90°$. Art: $\frac{18}{120} = 15\%$, $54°$. Reading: $\frac{24}{120} = 20\%$, $72°$. Check: $144 + 90 + 54 + 72 = 360°$.

28. A fair number cube has 6 equally likely outcomes, making it a uniform model.

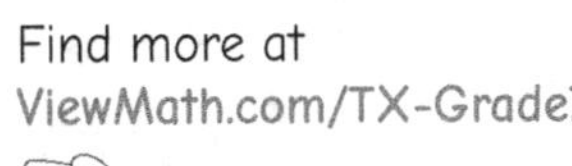

Find more at
ViewMath.com/TX-Grade7

29 $P(heads) = \frac{1}{2}$. Numbers greater than 4: $5, 6$, so $P = \frac{2}{6} = \frac{1}{3}$. $P = \frac{1}{2} \times \frac{1}{3} = \frac{1}{6}$.

30 A spinner with 4 equal sections has $\frac{1}{4} = 25\%$ chance for each section. Assigning one section as "win" gives 25%.

✅ Practice Test 8 — Answer Key

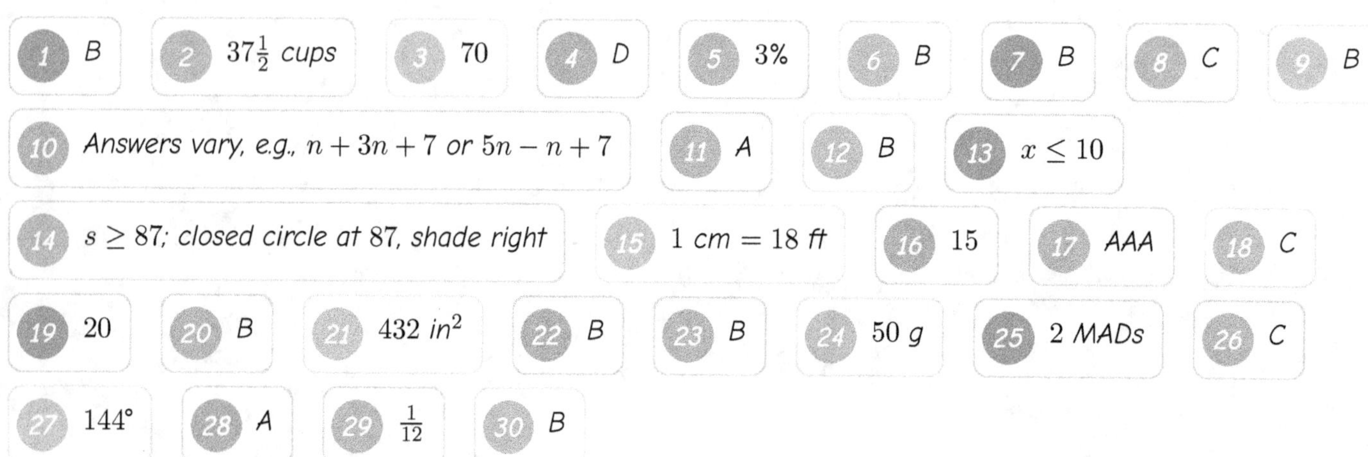

1 B	**2** $37\frac{1}{2}$ cups
3 70	**4** D
5 3%	**6** B
7 B	**8** C
9 B	

10 Answers vary, e.g., $n + 3n + 7$ or $5n - n + 7$ **11** A **12** B **13** $x \leq 10$

14 $s \geq 87$; closed circle at 87, shade right **15** $1\ cm = 18\ ft$ **16** 15 **17** AAA **18** C

19 20 **20** B **21** $432\ in^2$ **22** B **23** B **24** 50 g **25** 2 MADs **26** C

27 144° **28** A **29** $\frac{1}{12}$ **30** B

💡 Time to Learn! 💡

Review the explanations below, **especially for the questions you missed**.

Understanding why each answer is correct builds stronger problem-solving skills.

Tip: Circle any questions you got wrong, then read their explanation carefully.

📖 Practice Test 8 — Detailed Explanations

1 $k = \frac{150}{2.5} = 60\ miles\ per\ hour.$

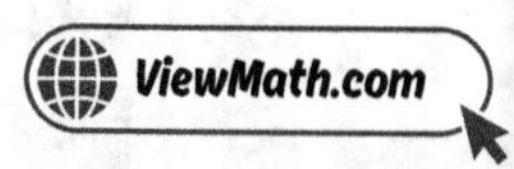

2. $k = \frac{7.5}{3} = 2.5$ cups per batch. For 15 batches: $2.5 \times 15 = 37.5 = 37\frac{1}{2}$ cups.

3. Divide: $56 \div 0.80 = 70$.

4. Change: $54 - 45 = 9$. Percent increase: $9 \div 45 = 0.20 = 20\%$.

5. $180 = 2{,}000 \times r \times 3 \ \Rightarrow\ 180 = 6{,}000r \ \Rightarrow\ r = 0.03 = 3\%$.

6. Bank Y earned \$35 on \$1,000, which is 3.5% — the highest of the three.

7. $200 - 60 - 80 = \$60$ available to save.

8. t is the number of years the money is invested or borrowed.

9. Substitute: $-4(3) + 10 = -12 + 10 = -2$.

10. Any expression whose like terms combine to give $4n + 7$ is correct. For example: $n + 3n + 7 = 4n + 7$.

11. Two identical groups of $(x + 4)$: $2(x + 4) = 22$. Divide by 2: $x + 4 = 11$. Subtract 4: $x = 7$.

12. Multiply every term by 6: $x - 4 = 3$. Add 4: $x = 7$. Check: $\frac{7}{6} - \frac{2}{3} = \frac{7}{6} - \frac{4}{6} = \frac{3}{6} = \frac{1}{2}$ ✓.

13. Subtract 7: $\frac{x}{2} \leq 5$. Multiply by 2: $x \leq 10$.

14. $\frac{78+85+90+s}{4} \geq 85$. Multiply by 4: $253 + s \geq 340$. Subtract 253: $s \geq 87$. Closed circle at 87, shade right.

15. Actual length: $9 \times 6 = 54$ ft. New scale: $54 \div 3 = 18$ ft per cm.

Find more at
ViewMath.com/TX-Grade7

16 The third side must be less than $7 + 9 = 16$. The largest whole number less than 16 is 15.

17 Three angles determine the shape but not the size. Infinitely many similar triangles can share the same three angle measures.

18 A cone has a circular base, and all horizontal cuts parallel to the base produce circles of varying sizes.

19 Scale factor: $\frac{12.5}{5} = 2.5$. $XY = 8 \times 2.5 = 20$.

20 Rectangle: $10 \times 4 = 40\ m^2$. Triangle: $\frac{1}{2} \times 10 \times 3 = 15\ m^2$. Total: $40 + 15 = 55\ m^2$.

21 $SA = 2(12 \times 8) + 2(12 \times 6) + 2(8 \times 6) = 192 + 144 + 96 = 432\ in^2$.

22 The general formula for the volume of any prism is $V = Bh$, where B is the area of the base and h is the height.

23 Selecting every fifth student from a list is systematic and gives all students an equal chance, making it more representative.

24 $\frac{49+51+50}{3} = \frac{150}{3} = 50\ g$.

25 Difference: $57 - 45 = 12$. In MADs: $\frac{12}{6} = 2$.

26 Count all the leaves: $3 + 5 + 2 = 10$ data values.

27 $\frac{24}{60} = 0.40 = 40\%$. Angle $= 0.40 \times 360° = 144°$.

28 Predicted: $\frac{5}{10} = 0.5$. Experimental: $\frac{24}{40} = 0.6$. The small difference is expected from natural variation.

29 $P(tails) = \frac{1}{2}$, $P(1) = \frac{1}{6}$. $P = \frac{1}{2} \times \frac{1}{6} = \frac{1}{12}$.

30 More trials produce more reliable estimates. The experimental probability approaches the theoretical value as trials increase.

✅ Practice Test 9 — Answer Key

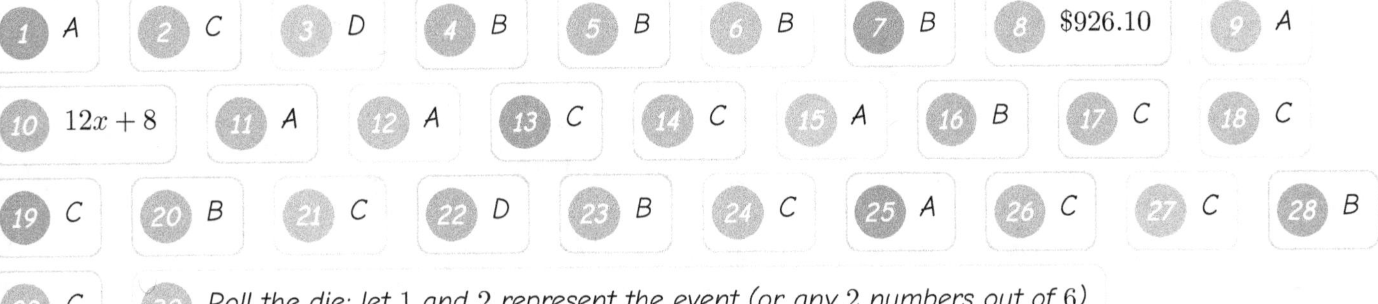

1	A	2	C	3	D	4	B	5	B	6	B	7	B	8	$926.10	9	A
10	$12x + 8$	11	A	12	A	13	C	14	C	15	A	16	B	17	C	18	C
19	C	20	B	21	C	22	D	23	B	24	C	25	A	26	C	27	C	28	B

29 C

30 Roll the die; let 1 and 2 represent the event (or any 2 numbers out of 6)

💡 Time to Learn! 💡

Review the explanations below, **especially for the questions you missed**.

Understanding why each answer is correct builds stronger problem-solving skills.

Tip: Circle any questions you got wrong, then read their explanation carefully.

📖 Practice Test 9 — Detailed Explanations

1 By convention, $k = \frac{y}{x} = \frac{20}{8} = 2.5$. Student B computed $\frac{x}{y}$, which is the reciprocal — not k.

2 Unit price: $42.50 \div 5 = 8.50$ each. For 8: $8.50 \times 8 = 68.00$.

Find more at
ViewMath.com/TX-Grade7

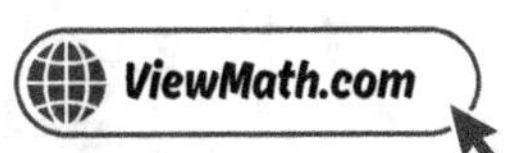

3. $40\% = 0.40.$ *Multiply:* $0.40 \times 150 = 60.$

4. *A 40% decrease means you keep* $100\% - 40\% = 60\%.$ *The multiplier is* $0.60.$

5. $120 = 1{,}500 \times 0.04 \times t \Rightarrow 120 = 60t \Rightarrow t = 2$ *years.*

6. *Interest* $= 800 \times 0.03 = \$24.$

7. *Expenses* $= 3{,}000 - 450 = \$2{,}550.$

8. *Year 3 interest:* $882 \times 0.05 = \$44.10.$ *Balance:* $882 + 44.10 = \$926.10.$

9. $5n$ *means "5 times a number." Subtracting 8 means "8 less than" that product.*

10. *Perimeter* $= 2(4x - 1) + 2(2x + 5) = 8x - 2 + 4x + 10 = 12x + 8.$

11. *Divide by 5:* $n + 3 = -2.$ *Subtract 3:* $n = -5.$ *Check:* $5(-5 + 3) = 5(-2) = -10$ ✓.

12. *Subtract 3:* $0.5x = 5.$ *Divide by 0.5:* $x = 10.$ *Check:* $0.5(10) + 3 = 5 + 3 = 8$ ✓.

13. *Solve:* $4x > 12,$ *so* $x > 3.$ *Only* $x = 4$ *satisfies* $x > 3.$ *Check:* $4(4) - 1 = 15 > 11$ ✓.

14. $x \leq -2$: *closed circle (includes* -2*) and shade left (values less than or equal to* -2*).*

15. *Scale ratio:* $\frac{25}{75} = \frac{1}{3}.$ *New length:* $6 \times \frac{1}{3} = 2$ *cm.*

16. *A square with area* 36 *cm*^{2} *has side 6 cm. All squares with side 6 cm are congruent, so exactly one square satisfies this.*

Find more at
ViewMath.com/TX-Grade7

17 $4 + 4 = 8 < 9$. The two shorter sides cannot reach across the longest side.

18 Every cross-section of a sphere is a circle. The size of the circle depends on where you cut, but the shape is always a circle.

19 Scale factor: $\frac{21}{7} = 3$. Area ratio: $3^2 = 9$. So the ratio is $1 : 9$.

20 Rectangle: $8 \times 6 = 48$ in^2. Two semicircles $=$ one full circle with $r = 3$: $3.14 \times 9 = 28.26$ in^2. Total: $48 + 28.26 = 76.26$ in^2.

21 $94 = 2(5 \times 3) + 2(5 \times h) + 2(3 \times h) = 30 + 10h + 6h = 30 + 16h$. So $16h = 64$, $h = 4$ cm.

22 $V = lwh = 6 \times 4 \times 3 = 72$ cm^3.

23 Testing every light bulb would destroy them all. Sampling lets you estimate the average without testing every one.

24 Average the three sample means: $\frac{78+81+80}{3} = \frac{239}{3} \approx 79.67$.

25 Player A has MAD $= 3$ (vs. 7) and range $= 10$ (vs. 25), so Player A's scores are more tightly clustered around the mean.

26 With an even number of values (12), the median is the average of the 6th and 7th values.

27 $\frac{3}{24} = 0.125 = 12.5\%$.

28 Predicted: $\frac{1}{4} = 0.25$. Observed: $\frac{25}{80} = 0.3125$. They are close, which supports the model.

29 Only one outcome out of 8 is all tails (TTT). $P = \frac{1}{8}$.

Find more at
ViewMath.com/TX-Grade7

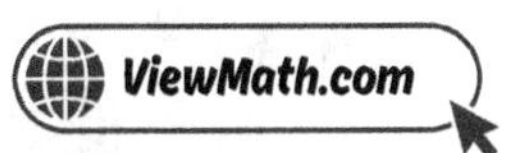

30 *Choose 2 of the 6 faces to represent the event.* $P = \frac{2}{6} = \frac{1}{3}$.

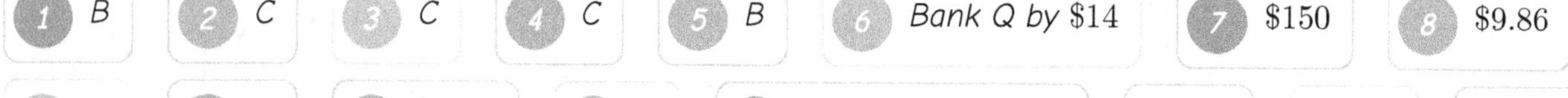

✅ Practice Test 10 — Answer Key

1 B	**2** C	**3** C	**4** C	**5** B
6 Bank Q by $14	**7** $150	**8** $9.86		

9 B **10** B **11** $x = 6$ **12** A **13** $2x + 8 > 20;\ x > 6$ **14** A **15** C **16** No

17 C **18** Rectangle; 8 cm by 9 cm **19** C **20** A **21** C **22** C

23 Only listeners who choose to call participate, so it is voluntary and biased **24** D **25** B **26** 34.5

27 B **28** B **29** B **30** A

💡 Time to Learn! 💡

Review the explanations below, **especially for the questions you missed**.

Understanding why each answer is correct builds stronger problem-solving skills.

Tip: Circle any questions you got wrong, then read their explanation carefully.

📖 Practice Test 10 — Detailed Explanations

1 Use any labeled point: $k = \frac{2}{3}$ from $(3, 2)$. Check: $\frac{4}{6} = \frac{2}{3}$ ✓.

2 Rate: $90 \div 2 = 45$ mph. In 9 hours: $45 \times 9 = 405$ miles.

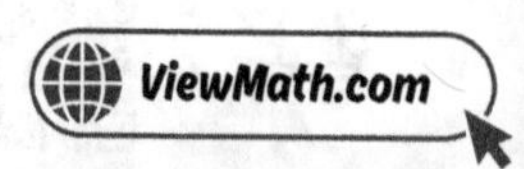

3   *Divide the part by the whole:* $18 \div 72 = 0.25 = 25\%$.

4   *A:* $100 \times 1.30 = 130$. *B:* $100 \times 1.20 \times 1.10 = 132$. *C:* $100 \times 1.15 \times 1.15 = 132.25$. *C is the greatest.*

5 $264 = 1{,}600 \times 0.055 \times t$ ⟹ $264 = 88t$ ⟹ $t = 3$ *years.*

6   *Bank P net:* $40 - 36 = \$4$ *gain. Bank Q net:* $30 - 12 = \$18$ *gain. Bank Q is better by* $18 - 4 = \$14$.

7 $1{,}800 \div 12 = \$150$ *per month.*

8   *Simple:* $I = 500 \times 0.08 \times 3 = \$120$, *total* $= \$620$. *Compound:* $A = 500(1.08)^3 = 500 \times 1.259712 = \$629.86$. *Difference:* $629.86 - 620 = \$9.86$.

9   *Substitute:* $2(-3) + 7 = -6 + 7 = 1$.

10   *Combine variable terms:* $3y - 5y = -2y$. *Combine constants:* $8 - 2 = 6$. *Result:* $-2y + 6$.

11   *Total area* $= 6 \times 2 \times (x + 1) = 84$. *So* $12(x + 1) = 84$. *Divide by 12:* $x + 1 = 7$. *Subtract 1:* $x = 6$.

12   *Add 2.4:* $0.6x = 3.6$. *Divide by 0.6:* $x = 6$. *Check:* $0.6(6) - 2.4 = 3.6 - 2.4 = 1.2$ ✓.

13   *The left side is heavier (tips down), so* $2x + 8 > 20$. *Subtract 8:* $2x > 12$. *Divide by 2:* $x > 6$.

14   *Subtract 6:* $-2x \leq -10$. *Divide by -2 and flip:* $x \geq 5$. *Closed circle at 5, shade right.*

15   *Actual road:* $7 \times 5 = 35$ *km. New scale:* $35 \div 3.5 = 10$. *So the second map uses* 1 *cm* $= 10$ *km.*

16 $6 + 3 = 9 < 10$. *The two shorter sides cannot span the longest side. The triangle inequality is not satisfied.*

Find more at
ViewMath.com/TX-Grade7

17 Although $90 + 100 + (-10) = 180$, every angle in a triangle must be greater than $0°$. A negative angle is impossible.

18 A vertical cut through the center of a cylinder produces a rectangle. Width = diameter = $2 \times 4 = 8$ cm. Height = 9 cm.

19 $\frac{60}{40} = \frac{h}{12}$. Cross-multiply: $40h = 720$. Divide: $h = 18$ ft.

20 Yard: $20 \times 15 = 300$ m^2. Circle: $3.14 \times 9 = 28.26$ m^2. Remaining: $300 - 28.26 = 271.74$ m^2.

21 A cube has 6 faces. $SA = 6s^2 = 6 \times 16 = 96$ cm^2.

22 $B = \frac{1}{2} \times 5 \times 12 = 30$ cm^2. $V = 30 \times 15 = 450$ cm^3.

23 Voluntary response samples are biased because people with strong opinions are more likely to respond.

24 $\frac{16}{40} = 40\%$. Then 40% of $600 = 240$.

25 Group A range $= 20$, Group B range $= 60$. Group B's data is far more spread out.

26 6 values: median $= \frac{33+36}{2} = \frac{69}{2} = 34.5$.

27 The entire circle represents the whole — 100% of the data or $360°$.

28 Red covers half the spinner ($180°$), while Blue and Green each cover a quarter ($90°$ each). This is a non-uniform model.

29 Pairs with product 12: $(2,6), (6,2), (3,4), (4,3)$ — that is 4 out of 36. $P = \frac{4}{36} = \frac{1}{9}$.

Find more at
ViewMath.com/TX-Grade7

 $P(sum = 7) = \frac{6}{36} = \frac{1}{6}$. Expected: $36 \times \frac{1}{6} = 6$. The simulation gave 8, slightly above expected.

Well done checking your answers!

Keep practicing to strengthen your skills.

Author's Final Note

I hope you enjoyed this book as much as I enjoyed writing it. Whether you are a student working through the material, a parent supporting your child's learning, or a teacher guiding your class, I have tried to make this book as clear and engaging as possible. I hope I have succeeded. If you have any suggestions for improvement, please let me know. I would love to hear from you.

The accuracy of calculations is very important to me. We have done our best, but I also expect that I have made some minor errors. Constant improvement is the name of the game. If you find any errors, please let me know. I will fix them in the next edition.

For students: Your learning journey does not end here. I have written a series of books to help you learn math. Make sure you browse through them. I especially recommend workbooks and practice tests to help you prepare for your exams.

For parents: Thank you for investing in your child's education. I encourage you to explore the companion resources available online to help support your child outside the classroom.

For teachers: Thank you for the invaluable work you do every day. I hope this book serves as a useful resource in your classroom. Feel free to reach out if you have suggestions or would like to discuss how best to use this book with your students.

I also enjoy reading your reviews. If you have a moment, please leave a review on where you found this book. It will help others find this book. If you have any questions or comments, please feel free to contact me at drNazari@ViewMath.com.

And one last thing: Remember to use online resources for additional help. I recommend using the resources on https://ViewMath.com You can find video lessons, practice problems, and more. You can also use the online companion for this book to track your progress and access additional resources.

Wishing all students the best in their studies, parents every success in supporting their children, and teachers continued inspiration in their classrooms!

Dr. A. Nazari

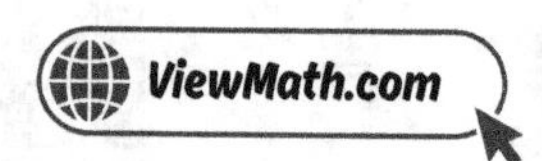

Great Job! Keep Learning with ViewMath!

*Keep up the great work! Visit **viewmath.com/TX-Grade7** for free lessons, quizzes, and more.*

Study Guide

Workbook

Step-by-Step

3 Practice Tests

5 Practice Tests

7 Practice Tests

Find more at
ViewMath.com/TX-Grade7

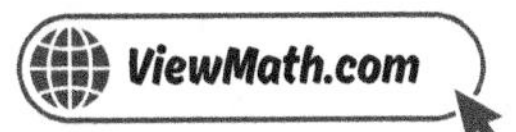